Everything can be taken from a man but one thing:
the last of the human freedoms—to choose one's attitude in any
given set of circumstances, to choose one's own way.

—Viktor E. Frankl
MAN'S SEARCH FOR MEANING (1959)

INCURABLE

Charles Harris

INCURABLE

A Life After Diagnosis

 COLD SPRING HARBOR LABORATORY PRESS
Cold Spring Harbor, New York · www.cshlpress.com

Incurable: A Life After Diagnosis
All rights reserved
© 2011 by Charles E. Harris
Published by Cold Spring Harbor Laboratory Press
Printed in the United States of America

Publisher and Acquisition Editor: JOHN INGLIS
Editors: MATTHEW STEVENSON and MICHAEL MARTIN
Design and Production: NANETTE STEVENSON
Cover Design: NANETTE STEVENSON
Back Cover Photo: MATTHEW STEVENSON

Front Cover Artwork:
A HORSE FRIGHTENED BY A LION
George Stubbs (1770)
© Walker Art Gallery, National Museums Liverpool

Library of Congress Cataloging-in-Publication Data

Harris, Charles, 1943-2010
  A life after diagnosis / Charles Harris.
      p. cm.
  Includes index.
  ISBN 978-1-936113-10-1 (hardcover : alk. paper)
  1. Harris, Charles, 1943---Blogs. 2. Colon
(Anatomy)--Cancer--Patients--United States--Blogs. 3. Liver
metastasis--Patients--United States--Blogs. I. Title.
  RC280.C6H37 2011
  362.196'9943470092--dc22
  [B]
                                    2010030067

All Cold Spring Harbor Laboratory Press publications may be ordered directly from Cold
Spring Harbor Laboratory Press, 500 Sunnyside Blvd., Woodbury, New York 11797-2924.
Phone: 1-800-843-4388 in Continental U.S. and Canada. All other locations: (516) 422-4100.
FAX: (516) 422-4097. E-mail: cshpress@cshl.edu. For a complete catalog of all Cold Spring
Harbor Laboratory Press publications, visit our website at http://www.cshlpress.com/.

10 9 8 7 6 5 4 3 2

For my family,
and friends

CONTENTS

FOREWORD

On March 2, 2009, thinking that the pain in my abdomen might indicate a failed hernia repair, I visited a surgeon in New York City. While sitting in his waiting room, my cell phone buzzed. The call was from my regular physician, a gastroenterologist, who reported that an MRI showed extensive cancer in my liver. As I was ending that brief conversation, the surgeon's nurse called me into the doctor's office. After examining me for less than a minute, he told me that I did not have a failed hernia repair; he thought that I had a tumor in my colon. I realized instantly that I must have colon cancer that had metastasized into my liver and that I did not have long to live.

I found myself trying to sort through the mundane logistical requirements of my situation—everything from whether I should keep my appointment in a few minutes with my dentist to the difficulties of passing on my bad news. I realized that communications would be an immediate and ongoing problem: whom to inform, by what medium, in what sequence? What about updating family and friends as I received new information? How many times could I stand to repeat the latest news about my illness? As I got sicker, how would I continue to communicate?

Several days later, a friend, Andrew W. Lo, solved the communications problem. Andrew suggested that I start a blog. I have always been a private person. When asked, "How are you?" my answer had always been, "Fine." Previously, the notion of describing my health on a blog would have been anathema. Now, it was the solution.

Eight days after my initial cancer diagnosis, I wrote my first post. At first, I gave no thought to using the blog for anything other than conveying news of test results and treatments to family and friends. While I was relieved not to have to repeat myself over and over on the telephone—especially when I was exhausted from a long day at the hospital—I came to realize that the blog was a godsend.

One of the saddest things that a terminally ill cancer patient can do is isolate himself. A blog strikes down one of the inherent barriers between an incurable cancer patient and the outside world: the elephant in the room. If a family member or a friend does not have up-to-date information about the patient's condition, no matter how lighthearted a conversation might be, the unspoken question of cancer hangs in the air. However, I have had to guard against the presumptuousness of assuming that everyone that I know had nothing better to do with their time than to read my blog.

In writing the blog, I found diversion from the cancer patient's endless self-maintenance chores. Writing not only gave me something to do, but also forced me to focus on my readers. I enjoyed getting lost in the crafting of words, phrases, sentences, and paragraphs; and in developing themes and twists and turns of narratives. Trying to write as well as I could was hard work, especially when I was tired from complications of treatments. Hard work demanded concentration, and concentration deepened diversion. The blog became talking therapy.

I cared about writing the blog as well as I could. When I have to read something that is sloppily written, which usually means that it contains unintended ambiguities, I get annoyed. I was mindful that my mother and father would be reading my blog. Former colleagues, whose writing I used to correct, would be reading it. Erudite friends would be reading it. What if they thought that I didn't know any better, didn't care anymore, or had chemo brain?

All themes, even if they concern mortality, cloy if unrelieved. After writing about my medical condition for some days, I began to be bored with the topics of death and dying, which surely meant that readers of my blog needed breaks from unrelenting, clinical dreariness. Because cancer lends itself to sadness, I began writing about other subjects. It would have been misleading anyway to give readers the impression that I was only thinking about my cancer or that I was constantly under treatment. Choosing topics for the blog from day to day seemed to me like mixing up pitches in a baseball game. If I threw a high, inside fastball, before coming back with more high heat, I needed over the next few days to toss some off-speed, gentler pitches.

In the aftermath of two major abdominal surgeries in the summer of 2009, I found that I had difficulty forming complex thoughts and could not hold them in my mind long enough to work with them. Until I trusted my brain again, I stopped posting on the blog. During those two periods, my wife, Susan, posted in my stead.

One of cancer's lessons for me is that living with knowledge of terminal illness does not relieve me of responsibility to others. On the contrary, I feel that my responsibilities to loved ones have increased. Certainly, I had no license to invade the privacy of living family members by writing about them. I did, however, write about my father in the context of people who have "perfect pitch" for their professions and about my Aunt Lucy, in the course of memorializing her daughter, my cousin Missy. Apprehending that the end was drawing near, I also wrote a blog entry, entitled "Susan," about my wife.

Finding meaning in the suffering and humiliations of cancer cannot be easy for anyone. As I am not a wishful thinker, the equipoise of certainty is unavailable to me. Other than as a catalyst, I find no inherent meaning in my cancer. In my life after diagnosis as an incurable, I am trying, in the face of growing horror, to become a better person and to celebrate life each day.

This book is no dirge. During the interlude between my diagnosis and death, much of my time has been happy and fulfilling, and I have continued to find unending comedy in the human condition. I have never lost my optimism about human progress. While I have accepted that I am beyond cure in the current primitive state of cancer therapeutics, I have never lost my faith in basic research's benefits for future cancer patients.

In my case, the details concerned being diagnosed as incurable; then adjusting my psyche to a twenty percent chance of cure through a combination of surgeries and chemotherapies; then resuming life as an incurable after having any hope dashed after the surgeries by a quick reappearance of cancer.

From the beginning of the blog, readers and I have known how this story will end. If there is any drama in my saga, it lies in the details of how one man, naked of illusion, reacted to, and thought about, living with incurable cancer.

C.E.H.
June 16, 2010, New York City

Acknowledgments

Matthew Stevenson convinced me to take the
entries from my blog and incorporate them into this book.
As editor and steward of the project, he brought on
Michael Martin to add his insight as co-editor, and
Nanette Stevenson to design the elegant jacket
and book, and to lead the production.
When John Inglis, Executive Director and Publisher of
Cold Spring Harbor Laboratory Press, learned of the project,
to my delight he asked to become the publisher of this book.
I appreciate the contributions of my literary executor,
Al Perry, especially his keen editorial eye.

—*Charles Harris*

I would like to thank the doctors, nurses, and staff at Memorial Sloan-Kettering Cancer Center for their skillful care, kindness, and empathy during this difficult time. Charlie especially appreciated that Dr. Leonard Saltz, his medical oncologist, listened to his desire to favor quality over quantity of life and helped him make decisions that allowed months of travel and enjoyment during the period after diagnosis; that Dr. William Jarnagin, his surgeon, made a heroic effort to give him a chance at a cure; and that Dr. Paul Glare, his palliative care specialist, worked with him to achieve dignity and enjoyment in his final days. We both developed a deep admiration for the nursing profession, and, in particular, for the nurses and nurse practitioners at MSKCC.

—*Susan Harris*

The editors wish to acknowledge Susan Harris
for her contribution to this book, in the posts and Postscript.

DIAGNOSIS

*To talk at all interestingly about death
is to talk about life.*

—D.J. Enright
THE OXFORD BOOK OF DEATH

DIAGNOSIS

I first heard the word *cancer* from my doctor last Monday, March 2.

On Thursday, March 5, I spent seven hours at Memorial Sloan-Kettering Cancer Center (MSKCC) undergoing various tests, including a CAT scan.

On Monday, March 9, I underwent a colonoscopy that confirmed that I have a tumor in the right colon that has metastasized to both lobes of the liver. As a result of the cancer, I am anemic.

On Thursday, March 12, Susan and I will have a first meeting and consultation with Dr. Leonard Saltz, the medical oncologist at MSKCC who will be directing my care.

I am deeply touched by and grateful for the many messages I have received from family and friends. It has been suggested to me that a good way to enable people to follow my news without having to go through the awkwardness of asking is for Susan and me to post such news on this blog. I can communicate with all of you about other things in normal, interactive fashion by e-mail, telephone, and in person. Although this blog is not set up for readers to post replies, please keep your communications coming through other channels. I find that e-mails are especially good, as I can read and respond to them at any time or place; even sitting, waiting for another test, at MSKCC. You are all very much in my thoughts.

Thank you,

Charlie

PHYSICAL SYMPTOMS

A few of you have asked if I am in any pain. When I move in certain ways, I have some internal pain in the location of the tumor. The area over the tumor is sensitive to even very light pressure. But none of the pain is bad enough to require painkillers, and I can usually get into a pain-free position when I want to sleep.

I am now anemic. I am told that I am pale, and I do seem to be losing strength on a daily basis. My only current exercise is going for short daily walks with my son, David. My weight has stabilized over the last few days as I have made a concerted effort to eat a lot of high-calorie food.

Overall, in terms of discomfort and loss of stamina, it feels as if I had hernia surgery a few days ago. Before I got my diagnosis, I thought that I must have torn an old hernia repair.

MEETING WITH MEDICAL ONCOLOGIST

As scheduled, Susan and I met today for the first time with Dr. Saltz. He confirmed that there are extensive metastases in my liver. It appears that there is lymph node involvement. Dr. Saltz told us that my cancer is incurable. We accept his opinion.

Dr. Saltz, however, recommends treatment. If successful, such treatment would extend my life by some indeterminable length of time, with periods in which I would be stronger than I am today and would be able to engage in an active life. Because my condition seems to be declining rapidly, he recommends that I begin chemotherapy tomorrow. He does not think that blood transfusions are called for at this time.

Dr. Saltz's plan includes my beginning with six to seven months of chemotherapy. He is starting with chemotherapy, rather than surgery on the colon, because during the recovery time from surgery, I would not be strong enough to receive the chemo I need for the liver. Accordingly, I will begin chemo as an outpatient tomorrow. The plan is for me to receive a cocktail, known by the acronym FOLFIRI, of irinotecan, leucovorin, and fluorouracil over a period of two days, with twelve days off, and then to repeat the treatment.

MARCH 13, 2009
FRIDAY 9:30 PM

FIRST DAY OF CHEMOTHERAPY

Several of you have been kind enough to e-mail or telephone to ask how my first day of chemotherapy went. I was concerned about my tolerance of toxic drugs, because I have historically had poor and atypical reactions to drugs in general and avoid them if possible. For example, one of my glaucoma medications causes, as a side effect, dilation of the pupils. In my case, this medication causes my pupils to contract, a reaction apparently not found in the literature until a research scientist documented the reaction in me.

It seems to me that the infusions went pretty well. They had to be stopped only once, to give me atropine when I started having an adverse reaction to one of the chemotherapies, irinotecan.

The pleasant surprise is that I got an almost immediate increase in energy and decrease in pain from the chemotherapy session, which I am guessing is the result of an infusion of iron and some steroid tablets that they gave me prior to the chemotherapy. Dr.

Saltz had expressed that it was his sense that I was on the brink of "sliding off the table," although I do not feel in imminent danger of precipitous decline as I write this tonight. I was able to take advantage of a pleasant winter's day by walking outdoors cumulatively for a good hour, with energy to spare.

MARCH 14, 2009
SATURDAY 10:33 PM

SIDE EFFECTS AND CHAMPAGNE

It has been explained to me that the side effects of chemotherapy generally grow worse over time, although a given session's side effects may be less severe than the previous session. I can't read too much into the side effects that I have experienced during the day and a half since the end of my first infusions. Nevertheless, I am pleased to report that I have found initial side effects to be tolerable. Until sometime tomorrow, I will continue to be hooked to a portable bottle of fluorouracil. I am told this drug can change taste buds, but we opened a bottle of rose Champagne tonight, and I found it delightful. I've been told to exercise as much as possible, so I walked to the Central Park Zoo and back on a pleasant family outing that lasted almost two hours.

MARCH 15, 2009
SUNDAY 9:18 PM

LITTLE APPETITE

This morning, I felt energetic. Later in the day, I began to have migraine attacks, a total of three altogether. I hope that the chemotherapy is not going to trigger such attacks routinely. In any event, whether as a consequence of the chemotherapy or the migraines, I have lost much of my appetite.

If all goes according to form, tomorrow and Tuesday should be the nadir of this chemotherapy cycle. I am planning on seeing friends starting on Thursday, so please do not hesitate to get in touch if you would like to get together. I would love to see you.

MARCH 16, 2009
MONDAY 10:25 PM

TO AN INDIAN RESTAURANT

Today's predominant side effect was persistent, moderate nausea. As I have been instructed to try to eat, Susan suggested that spicy Indian food might pique my appetite. We went out to dinner at a favorite local restaurant, and she was proven correct. I even had a glass of Gewürztraminer with dinner.

MARCH 17, 2009
TUESDAY 8:41 PM

FEELING BETTER

I hadn't planned to try to get together with any friends until Thursday, but I woke up feeling better today than yesterday. So I called a friend, and we walked to a simple, neighborhood, Italian restaurant for lunch. My seat faced the window, and I had the treat of watching the revelers in their St. Patrick's Day green delighting in the cool, sparkling sunshine.

FEVER

Today, I briefly developed a mild fever and a couple of other symptoms that might be attributable to reactions to chemotherapy. Earlier in the day, I had a pleasant lunch with old colleagues from Harris & Harris Group.

A FRIEND VISITS

My temperature was normal today. I had a visit from a dear old friend.

VIKTOR FRANKL

Among e-mails I received today from friends, I have selected two to share with you. One is an e-mail from a friend who is coming to see me on Monday. He writes, in response to last night's post in which I referred to a "dear old friend":

"Charlie," he wrote. "When you blog Monday evening, could you please refer to me as a 'dear' friend and 86 the 'old'?" Lesson learned. Henceforth, I'll try to refrain from characterizing friends in my blog.

A second e-mail read: "Charlie, tonight I read a quote from Viktor Frankl: 'Everything can be taken away from man but one thing: the last of human freedoms—to choose one's attitude in any given set of circumstances, to choose one's own way.' "

I replied by e-mail: "That is very insightful of you. When I found out I had cancer, I thought almost immediately of Frankl's *Man's Search for Meaning*, which I have reread periodically since I read it at least fifteen years ago. I realized that I had to find meaning in light of my new circumstances. When I got home, I got a copy from my bookshelves and reread several parts, including the passage that you quote. I do not think I have figured out 'meaning' yet, and I do not know if I will be able to do so, but Frankl is definitely helping me to frame the questions in my mind."

MARCH 21, 2009
SATURDAY 6:21 PM

THE LONELINESS OF DYING

Many, probably most, of you know from your personal experience much more about the realities of mortal illness than I do. I have the good fortune of still having my parents with me. My father's ninety-ninth birthday is April 4, and my mother's ninety-fifth is not far away. We recently celebrated their seventieth wedding anniversary.

I would like to share excerpts from an e-mail from a friend who wrote me today upon learning of my condition. "Having been through my mom's fatal illness last year, I can tell you that I deeply feel for you, your family and…your friends….Everyone may be different, but one thing I've learned is that most people with serious illnesses hunger for companionship and connection. They want to know and feel that they really matter to others."

I can attest that I am no exception. One of my first instincts upon learning of my condition was to fear loneliness and isolation and to seek the comfort of family and friends with a special intensity—an intensity that I suppose is born, in part, of the realization that I don't know how much time is left—not only, in my friend's words, to feel that I really matter to others, but also to let them know that they matter to me. Whatever the source of the intensity may be, one thing that I am sure of is that I want to fight as hard as I can fight against the fearful loneliness of dying.

MARCH 22, 2009

SUNDAY 6:14 PM

HORSES IN TRAINING

At lunch today with some friends from Boston, we discussed, among other things, horses. Susan and I currently have five thoroughbred racehorses in training. Four of them are in Florida, and one is in France. They are young horses, with their racing careers ahead of them. They include a four-year-old gelding, W.C. Jones; a three-year-old gelding, Hot Money; a three-year-old ridgling, Mustang Island; a two-year-old colt, Backslider; and a three-year-old filly, Sacred Music.

W.C. Jones has made three starts, finishing a fast-closing second in his most recent race on June 22, 2008, at Belmont Park, at a mile and an eighth on the turf. Since then, he has been recovering from an injury, first in Kentucky and currently at a training center in Florida. When he is ready to run, he will go to our trainer, Christophe Clement, in New York. I named W.C. Jones for my late uncle, who bred and raced thoroughbreds and got me interested in them in 1961. During the summer of 1963, between my junior and senior years at Princeton, I worked for his trainer at Saratoga racetrack. I bought my first thoroughbred racehorse in 1974, and since then, from time to time, Susan and I have owned and bred varying numbers of thoroughbreds—

sometimes in partnership with others, especially in recent years with the late Peter Karches and his partner, Mike Rankowitz.

Hot Money has made one start, at Aqueduct Racetrack on Long Island. We started him in a race on the dirt main track, just to give him experience. We think that he may prefer to run on the turf. We'll find out. He may be ready to run again in about a month, in New York.

Mustang Island was also given one race at Aqueduct before being put away for the winter. He may be ready to run again in six weeks or so, also in New York.

Backslider is at the stage of training in which owners and trainers find out whether or not a young horse has speed. A big part of managing racehorses is figuring out what they do best—sprint, compete at longer distances, take the lead early, come from behind, run on the dirt or turf, and so on. But if a horse is going to be top class doing anything, it has to have a "good turn of foot," i.e., the ability to run a fast quarter of a mile at some point in a race.

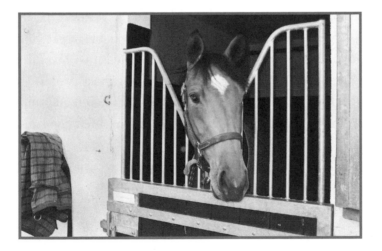

Sacred Music is in training in Chantilly, France, with trainer Nicolas Clement. All of our horses in the United States are currently trained by his brother, Christophe. Sacred Music is being pointed for her first race at one of the Parisian tracks in mid-April. She appears to be best suited to running long distances on the turf, which is the surface on which the French races are contested. Although there are many good turf races in the United States, most racing here is conducted on dirt or artificial surfaces.

Susan and I bred Hot Money, and I bought the other four horses at yearling auctions in Saratoga. Although, as you would expect, I recently gave some consideration to selling our horses, Susan and I have decided to keep them. Like all owners with young, largely untested horses at the beginnings of their careers, we are keenly following their progress as they are being prepared for this season's racing.

MARCH 23, 2009
MONDAY 8:38 PM

A BUSY DAY

This morning, I had more energy than I can recall having in the last few weeks. As a result, I may have gotten a bit too ambitious: I wound up having lunch with one friend, a visit with another in mid-afternoon, and dinner with another. (I must confess that I let myself get talked out of the check at both lunch and dinner.) Other than getting a migraine about 4:00 pm, I felt good all day.

TAKING CARE OF BUSINESS

This morning, Susan and I sat with lawyers while they witnessed the signing of my will, my living will, and a power of attorney. Our next order of business will be to consider retaining investment-advisory firms or a trust company.

At lunch, a friend treated me to Southern fried chicken, mashed potatoes, biscuits, and gravy at the University Club. Although I'm supposed to pack away all the calories I can, that lunch may have tested the limits of the theory.

LIFELINE

Much of my day was spent undergoing a scheduled, minor surgical procedure to install a Mediport. A Mediport is a device implanted under the skin of the chest that enables infusion of chemotherapy.

INFINITESIMAL CHANCE

Yesterday, before signing the consent to having the Mediport installed, I was told by the interventional radiologist who would perform the procedure that there was "an infinitesimal chance" of infection. I got infected. I was given intravenous antibiotics today for the infection and will not be given chemotherapy tomorrow as previously scheduled.

TWELVE-HOUR DAY

We left home at 8:15 a.m. to go to the hospital to have the infection caused by the implantation of the Mediport checked. At 1:30 p.m., we went back to the hospital when the infection appeared to worsen. At 8:20 p.m., we arrived back home. In the interim, the Mediport was removed; implantation of a PICC (peripherally inserted central catheter) through my right arm was attempted unsuccessfully because of thrombosis caused by implantation on March 12 and removal on March 15 of a PICC through the same arm; and a PICC was subsequently implanted through my left arm, where it now resides. A PICC, like a Mediport, enables infusion of chemotherapy.

I opted to undergo these procedures under local rather than general anesthesia on the understanding that, if I did so, I would have to stay in the recovery room for only an hour. But bureaucracy prevailed, the computerized instructions to the recovery room were botched, and the nurse wouldn't let me go home for the better part of three hours.

QUIET SATURDAY

Aside from a brief, scheduled trip to the hospital for some follow-up care, we spent a quiet day at home. I was tired from yesterday, so I just read a bit, and handicapped and watched some horse races on television with David. (We might have seen two really good contenders for the Kentucky Derby in today's first two finishers in the Florida Derby, Quality Road and Dunkirk.)

SHORT WALK

I had a little more strength today and went for a short walk with my son—0.68 miles, according to the pedometer that a friend sent me. We also watched Tiger Woods win the Bay Hill Invitational golf tournament. I was a little ambivalent about rooting for Woods after he threw a golf club three times. But when one invests the time to watch a golf tournament on television, it is hard to complain about such a payoff—getting to see another dramatic victory by what must be one of the greatest clutch performers in the history of sports.

TAKING CHANCES

Today, privately held BioVex, Inc., announced that it had raised $40,000,000 in a first close of a Series F round of financing. BioVex will use the proceeds of this financing to fund a Phase III study in metastatic melanoma with its lead cancer product, OncoVEX GM-CSF.

One of the previous investors in BioVex that participated in this financing was Harris & Harris Group, the firm that I was associated with from 1984 until 2008. I always felt privileged to get to spend the final years of my career making the sorts of investments that Harris & Harris Group makes. As an early-stage, venture-capital firm, Harris & Harris Group helps to develop, and puts capital behind, worthy ideas, intellectual property, and people.

Because I lack both proclivity for, and training in, science and technology, I never thought that I was qualified to be a venture capitalist. Nevertheless, once I had surrounded myself with people who were fully qualified, I had a useful role to play, because I was willing to take chances on people and projects that were not deemed fundable at the time by my peers at other firms.

There are a number of companies in Harris & Harris Group's current portfolio of approximately thirty privately held, risky companies that could, if they reach their potential, make a difference in the lives of people (as well as make a lot of money for Harris & Harris Group and its shareholders). Harris & Harris Group has actually played a much more central role in many of these companies than it has in BioVex, and BioVex is not its only portfolio company developing cancer therapeutics. But imagine if just BioVex, among Harris & Harris Group's portfolio of companies, reaches its potential and brings to market a cancer vaccine. Thinking of the lives that may be saved, and the suffering that may be averted, it seems to me that my entire career was more than well spent if all I accomplished was being in position to aid in development of a safe and efficacious vaccine.

MARCH 31, 2009
TUESDAY 2:19 PM

DOWN FOR THE COUNT

This morning, I was scheduled to receive the chemotherapy that was originally scheduled for last Friday, but my white blood cell count proved to be too low for chemotherapy. Instead, I received an injection of Neupogen and am scheduled to receive another tomorrow, with the goal of increasing my white blood cell count enough for me to receive chemotherapy on Thursday.

YOUNG HORSES IN
SPRINGTIME

The world is so exquisite, with so much love and moral depth,
that there is no reason to deceive ourselves with pretty stories
for which there's little good evidence. Far better, it seems to me,
in our vulnerability, is to look Death in the eye and
to be grateful every day for the brief but
magnificent opportunity that life provides.

—Carl Sagan
BILLIONS & BILLIONS

PERFECT PITCH

Last Monday, Susan and I were dinner guests of neighbors, Connie and Randy Jones. Among other things, we discussed Randy's new book, *The Richest Man in Town*. To write this book, Randy interviewed the richest self-made person in each of the hundred largest cities and towns in the United States. I asked whether Randy could discern any common traits in these people. One of the qualities that Randy noted was that they all seem to have, like musicians for music, "perfect pitch" for their enterprises.

In thinking about it, my father has perfect pitch for architecture. He started practicing architecture in 1932, started his own practice in 1935, and retired in 1997. He certainly has other interests (for example, he played golf or hit golf balls practically every day). But he was always content to go to his office on weekends to work on a project.

I'm not sure that I ever had perfect pitch for any of my businesses. I saw opportunities, figured out what would be required to capitalize on the opportunity, and assembled the requisite people, capital, and other resources. These were not businesses for which I was qualified or in which I was deeply interested. For example, Rick Childress and I started and managed a fund to invest in undervalued, publicly traded, industrial companies, and, later, a fund to invest in regional banks just before it became legal for banks to operate across state lines. Both funds were successful, and Rick and I had a lot of fun creating and managing them. Rick had the perfect pitch for the research underlying these funds that made them special.

When I saw that there was a shortage of property-and-casualty reinsurance capacity, I decided to start a new, publicly traded reinsurance company. ReCapital Corporation, of which I was chairman, was successful, but I had no interest in the insurance industry. The people that I brought in to run the company, led by the late Don Chisolm and by Jim Roberts, were the consummate professionals who had perfect pitch for reinsurance.

As I noted about BioVex, even though I am interested in venture capital, I regard the people that I brought into Harris & Harris Group as the ones with perfect pitch for venture capital based on their deep scientific and technological knowledge.

As for myself, I think the best that I could say is that I created opportunities for others who had perfect pitch for particular businesses.

APRIL 2, 2009
THURSDAY 3:00 PM

$5,000 A POP

Biotechnology is a wonderful thing. After two shots of Neupogen, within forty-eight hours, my white blood cell count went from 1.9 to 10. So, after a six-day delay, I get to have chemotherapy today. Henceforth, part of my regular protocol will be shots of Neulasta (at $5,000 a pop), a longer-acting form of Neupogen. Bone pain is a common side effect of Neupogen and Neulasta. I did experience the bone pain, but it responded amazingly to Tylenol.

APRIL 3, 2009
FRIDAY 5:39 PM

SACRED MUSIC

Sacred Music is entered to make her debut this Sunday, April 5, in the second race at the Parilly Racecourse in Lyon, France. The race is called the Prix Louis Saulnier, with a purse of 14,000 euros. It will be contested at 2,200 meters (1-3/8 miles) on turf and restricted to three-year-old fillies which have never won a race. Sacred Music drew the outside post position, number 11, and will be ridden by Jean-Michel Breux. As nine of the other fillies have run previously, Sacred Music will be conceding experience to most of the field, and we should probably be happy if she shows promise and comes back sound. The French do a good job of presenting their racing on the Internet, and Susan and I are trying to figure out if it is possible to watch the race live.

I walked about five miles today, mainly running errands, at a somewhat slower pace than I hope Sacred Music maintains on Sunday. One of my errands was to bring in for servicing the watch that my mother gave me for my college graduation in 1964. (I skipped the graduation ceremony to attend the Belmont Stakes.) The Rolex representative reminded me somewhat sternly that I had not had it serviced since 1999. What really hurt my feelings was that he informed me that the servicing would take longer because the watch is now considered to be an antique. Does that say anything about its owner?

APRIL 4, 2009
SATURDAY 3:34 PM

HAPPY 99TH BIRTHDAY, DAD

As my treatments precluded my traveling to Ponte Vedra, Florida, to celebrate my father's ninety-ninth birthday, I just wished him a happy birthday by telephone. Both he and my mother, who will turn ninety-five in July, are as sharp as ever—which is plenty sharp.

FIFTH IN DEBUT

According to trainer Nicolas Clement's description of Sacred Music's race, she ran about as expected. (We will receive a video of the race later, but we were not able to view it on the Internet.) Apparently, she broke somewhat slowly, was a bit shy with the other horses, and then finished fairly well, beaten about four lengths by the winner. Assuming she comes back in good order, Nicolas expects her to benefit from the race.

ADVENTURE CAPITALISM

Former colleagues in various ventures have commented about Randy Jones's concept of perfect pitch. During my years at Harris & Harris Group, we invested in eighty-four ventures. In the successful ventures, the venture capitalists like Harris & Harris Group play a role, but the requisite perfect pitch has to be within the management of each company.

In my earliest venture-capital deals, we not only started with our own modest capital, but also I might serve as the chief executive officer of that startup until I could recruit someone to run it who had what Randy Jones might term perfect pitch for the business. Sometimes, the recruited chief executives came from non-obvious backgrounds: They were in the process of perfecting their pitch.

Over a quarter of a century ago, prior to my days at Harris & Harris Group, my first venture-capital deal was Thoroughbred Equity Company (TECO). I started TECO as a joint venture with

Fasig-Tipton, one of the two major thoroughbred-horse auction houses. We formed TECO to enable breeders of thoroughbreds to use their bloodstock as financial collateral. Up until the time that I sold my interest in TECO to Fasig-Tipton for a high percentage return on my modest investment, TECO never had a loan loss. This impeccable record partially reflected my simple bonus policy: If there were any loss on a loan during the year, no one in the company would get a bonus at year end. But the people that I recruited to run the business were responsible for its success. First, I recruited Michael Lischin, a young lawyer from a small bank, with no horse-industry background, to run the business. Today, he is a successful thoroughbred breeder in New York.

Valuation of the collateral was a critical input into TECO's lending business. Rob Whiteley, a professor of psychology at Rutgers University who was passionate about thoroughbreds, which he bred on a modest scale at his small New Jersey farm, was a friend of mine. In as radical a career gamble as I can recall, Rob started appraising thoroughbreds for Fasig-Tipton and TECO. Later, the investor Carl Icahn wanted to enter the thoroughbred business and asked me to recommend someone to run the business for him. I recommended Rob, who was still relatively new to the thoroughbred business, on the basis of his brains and his unquestionable integrity. Rob built a successful breeding operation for Icahn. Today, Rob has his own farm in Kentucky, where he is a prominent breeder respected not only for his business prowess, but also for his sterling character and universal kindness to others. Rob has perfect pitch for his business and for life.

JULY IN APRIL

Susan and I just returned from a much-anticipated dinner with five other friends, aka the "Gang of Seven," hosted by one of the other couples at their apartment. Most years, we have gotten together as a group for the Fourth of July weekend, at the country place in Connecticut of one of the other couples. We discuss books and foreign policy, gossip a bit, and enjoy each other's company. Good food and wine stimulate conversation; long walks and rowing a boat on the lake stimulate appetite and thirst. Upon parting tonight, we began formulating plans for our next get-together.

ON THE BEACH

One of the books I've been thinking about lately is Nevil Shute's *On the Beach*. What struck me when I originally read it, and what has stuck in my mind ever since, is the behavior of the characters in this fictional work as they await in Australia the inexorable approach of fatal radiation poisoning from atomic warfare in the Northern Hemisphere. They go about their lives and mundane activities, planting gardens, and the like, as if they have all the time in the world. "Surely," I thought as a young man, "people would not behave that way—I certainly would not behave that way in the face of certain doom. I would devote myself to profound thoughts and extraordinary actions."

I now think Shute had it right. Starting tomorrow, I'm hoping to watch all of the televised portions of the Masters golf tournament. Because of work, I never watched all of the broadcast on

Thursday and Friday. Out of a feeling that I ought to be doing something more physically or mentally active, I don't know if I ever watched all of it on Saturday and Sunday either. I feel few such scruples this year.

LOCATION, LOCATION, LOCATION

When we bought our current apartment in 2003, like anyone else choosing a new residence, we put a lot of emphasis on location. We are located across from an entrance ramp to the Queensboro Bridge, which means that we can travel to the Long Island airports, golf courses, and racetracks without having to deal with Manhattan traffic. I had a twenty-minute walk to my midtown office, and we have similar walks to Carnegie Hall, Central Park, and much of midtown Manhattan. A crosstown bus stops almost in front of our building, and two subways have stops within less than ten-minute walks from our building. There is a kaleidoscope of ethnic restaurants within a ten-minute walk.

What we failed to appreciate is that we are located only twelve blocks from the main campus, and a mere six blocks from the outpatient chemotherapy facility, at MSKCC. We (either Susan or David usually goes with me) have been averaging three or four trips a week to the MSKCC facilities. Today, I had a scheduled visit to one of the facilities; last night, I had an unscheduled visit to the other. I really feel for the patients and their families who have to add difficult commutes to their burdens of disease and treatment.

GOOD FRIDAY

Yesterday was Good Friday, and I did indeed have a good day. I felt better than I have in a while and was able to enjoy a leisurely lunch with a friend at a superb, Greek seafood restaurant that offers a three-course, specially priced lunch menu.

A friend, who has had to endure a lot of physical pain, recently included in an e-mail to me his favorite expression, "Every day is good." Having found lots of references to "Every day is a good day," but none to his favorite expression, I asked him if he had originated it. He replied that it is too profound to be original—it is simply his favorite expression.

APRIL I3, 2009

MONDAY 6:19 PM

YOUNG HORSES IN SPRINGTIME

Easter, to me, has always been the real beginning of spring. Young horses blossom in the spring as they begin to acquire their summer coats. For many years, we have shipped our young horses to Florida for the winter, to prepare for their spring campaigns in New York.

Hot Money, the three-year-old chestnut gelding pictured above, should be ready to run in New York in a few weeks. Mustang Island, the three-year-old bay ridgling, is two or three weeks behind Hot Money in his preparation.

APRIL 14, 2009

TUESDAY 4:17 PM

THE TOOTH FAIRY

Following what I had envisioned as a routine 10:45 a.m. periodontist's appointment, I had scheduled lunch around the corner at Petrossian (which offers a superb three-course luncheon special) with a friend at noon. I kept my periodontist's appointment, but not my luncheon date. After the periodontist discovered that I had a tooth with two cracked roots, I settled for a tooth extraction at noon instead of the smoked salmon at Petrossian.

When one is undergoing chemotherapy, routine procedures become complicated and have to be coordinated with one's oncologist. If all goes well, I am still on track on Thursday to undergo, for the second time, a procedure to have a Mediport installed, and on Friday, to undergo a third round of chemotherapy. With these medical things to worry me, I forgot to bring the tooth home to put it under my pillow.

$24,303.80

My March bill for hospital charges from MSKCC arrived today, in the amount of $24,303.80. This bill does not cover a full month, as I had not yet been diagnosed with cancer at the beginning of March (although it seems to me that I received the news of my diagnosis a long time ago). Nor does this bill cover doctors' charges or prescriptions. Moreover, I was not receiving Neupogen and Neulasta in March. My current regimen includes fortnightly shots of Neulasta. Given how low my white blood cell count dropped after my first chemotherapy, I wonder if my treatment could have continued without Neupogen or Neulasta. Given the infection that was generated by the fractured molar that I had removed yesterday, I wonder what would have happened if my white blood cell count had still been 1.9, as it was before I received my first shot of Neupogen.

Because I am sixty-six years old, the lion's share of the cost of my treatment will be reimbursed by Medicare, which is good for me, but quite a burden on other current and future taxpayers. As medical science succeeds in making more expensive new treatments like Neulasta available, it will be increasingly difficult for us as a society to put off facing the excruciating ethical questions posed by the costs of someone like me with an incurable illness who has passed his years of peak productivity. Our economy's recent massive loss of wealth is going to make it that much more difficult to avoid dealing with the ethical questions that must be resolved in order to ration healthcare in a world of finite resources.

BELT AND SUSPENDERS

Susan and I arrived at the main campus of MSKCC today at 7:40 a.m. and left at 1:45 p.m. Once again, I am equipped with a Mediport. This time, the surgeon left the PICC in my left arm. Thus, if my white blood cell count is high enough tomorrow for me to receive chemotherapy as scheduled, the oncologists will probably utilize the PICC for the infusion, but could also utilize the Mediport. If no problems develop this time with the Mediport, the plan is to remove the PICC on Sunday, when I am scheduled to be unhooked from a forty-six-hour infusion of 5-FU and to be injected with Neulasta.

YESTERDAY ENDED THIS MORNING

Yesterday ended this morning at about 1:30 a.m. Just before dinner, at about 7:00 p.m., I noticed that the skin around the area where the Mediport is installed had turned red. The elapsed time from installation to first symptom and the area and degree of redness were roughly similar to those I had noticed when the first Mediport was installed and had to be removed. If this second Mediport were to be removed, I am not sure that the doctors would want to try again. The interventional radiologist who installed the latest one did so on the left side of my chest after viewing the still-healing site of the first attempted installation on the right side of my chest. (Before I had signed the consent for this second attempt, I had been told that the chances of infection were less than one percent for a Mediport installation.)

I went to the Urgent Care Center at MSKCC, an emergency room, where the interventional radiologist on call drew lines with a Sharpie around the area of reddened skin so that any changes in the progress of the infection could be noted in the morning. I went home, where I was greeted with the message that I had to call the Urgent Care Center immediately, which I did. The Urgent Care Center nurse who had performed my triage informed me that my blood pressure was 220 over 90 and that I had to come back to be evaluated by another doctor. As I have had "perfect" to "borderline low" blood pressure all of my life, I was incredulous; but realizing that now that I have cancer anything is possible, I reported back to the Urgent Care Center, where I was seen instantly. The nurse rechecked my blood pressure and got a similarly high reading. Both times, the nurse had taken the reading from a cuff around my leg, as I have a PICC in my left arm and phlebitis in my right arm from the prior installation and removal of a PICC. Faced with the prospect of having to be treated for high blood pressure on top of everything else, it occurred to me to request that, phlebitis or no phlebitis, the nurse take my blood pressure from a cuff around my right arm. That blood pressure reading was 135 over 79—high for me historically, perhaps because of the stress of the day's intervention and events, but no cause for alarm. (This morning, when I went for chemotherapy, my blood pressure was down to 107 over 69.) After an atypically short wait in Urgent Care, I was seen by a physician and discharged.

Remaining apprehensive about the progression of the redness around the site of the Mediport, I slept fitfully and then reexamined the site around 1:30 in the morning. The redness was receding. I went to the kitchen, raided the refrigerator, and went back to bed.

TURN OF MIND

It has become obvious that this blog has drifted away from just being the "latest news" about my condition. Although I keep a diary for the doctors of all of the changes in my condition, symptoms, side effects of chemotherapy, and the like, such detailed information seems inappropriate in a record for family and friends. Also, I think that most people have a good idea of what chemotherapy entails. Unless some facet of my treatment or regimen seems noteworthy, I haven't been including it in my blog.

I have less of an excuse for the other topics I have been addressing in this blog. So far, I have found that writing something in it each day has been therapeutic for me. I suppose that the exercise is both a form of talking cure and a little job for me to do each day. Talking cures are always good. And, having worked all my life until retiring on December 31, 2008, maybe I don't feel quite right unless I do a little work each day.

HITTING THE TRIFECTA

Because I hit a treatment trifecta yesterday, I decided to write for a second time today. Yesterday, the area around the Mediport looked increasingly healthy, not only to Susan's and my eyes, but also to the doctors and nurses. Because my white blood cell count was in the normal range, my chemotherapy proceeded as scheduled. The oncologists agreed to delay the infusion of iron sucrose, which treats my anemia, until two days after the infusion of my chemotherapy.

Last time I had chemotherapy, there was a similar two-day delay on the infusion of iron sucrose, because my doctor or his nurse had forgotten to order it. Because I subsequently had fewer and less severe side effects from the chemotherapy, I asked my doctor and nurse practitioner to repeat the delay in the infusion of the iron sucrose. They agreed, even though they think that my better experience with chemotherapy, after delaying the iron sucrose, was probably just a coincidence. But they accepted my theory that it will be easier for me to accept more adverse side affects with equanimity, knowing that at least we tried to capitalize on serendipity.

APRIL 19, 2009
SUNDAY 3:49 PM

SUNDAY CHORES

Three minor procedures were scheduled for me today—unhooking of the forty-six-hour infusion of 5-FU, an injection of Neulasta, and removal of the PICC in my left arm. Because today is Sunday, these procedures were performed in the Urgent Care Center, which entailed two-and-a-half hours of waiting before and between procedures. If all goes well, I will not report back to MSKCC at the chemotherapy facility for infusion of iron sucrose until 9:00 a.m. tomorrow. How do patients and their families cope with long commutes to their treatments, especially when they have interminable waits in the unpleasant environment of the Urgent Care Center?

RASH

Last evening, I developed a rash. The cancer patient's standard list of phenomena requiring an immediate call to his or her oncologist includes observation of a rash. My oncologist's covering physician listened by telephone to my description of the rash and decided that, unless it worsened, I could treat it with Benadryl until my previously scheduled appointment for infusion of iron sucrose at MSKCC's chemotherapy center at 9:00 a.m. this morning. The registered nurse who examined the rash this morning could not determine its cause or predict its likely course. Per recommendation, I have continued to take Benadryl, which has combined with last Friday's chemotherapy to tire me. For the first time since my diagnosis, I spent time in bed in the middle of the day.

I am apprehensive about the side effects of chemotherapy. My experience so far is that they manifest themselves suddenly and develop unpredictably. Although my primary worry about them is that they will limit my treatment, I also worry about how sick they will make me and to what extent they will limit my ability to be with friends.

MAKING THE ROUNDS

At 3:00 p.m. today, I had no medical appointments scheduled for tomorrow. As of 5:30 p.m., I have one with an oncologist, one with a dermatologist, one with a radiologist, and one with a liver surgeon. It should be a busy day.

"FORTY TEARS' WAR"

When I first typed the title of this blog (I'm a lousy typist), I mistyped it, "Forty Tears' War."

Starting with an article on the front page of today's *New York Times* (by Gina Kolata), the paper will publish a series of articles about "the struggle to defeat cancer." In today's articles, Ms. Kolata documents that the death rate from cancer has barely declined since 1950. She states that patients with colorectal cancer that has metastasized—and mine has metastasized significantly—have a five-year survival rate of less than ten percent. (Elsewhere, I have read that patients with my diagnosis have a mean survival rate of one to two years, with a two percent chance of surviving for five years if they are fifty-five years old.) The first article also quotes Dr. Saltz, who is treating me.

Dr. Saltz does not get drawn into statistical discussions with me about how long I may live, pointing out that no patient is a statistical average, that the data are not very good, and that the bell-shaped curve that would apply is flat. I might already be part of the statistics if my cracked tooth and abscessed gum of last week had coincided with my chemotherapy-depressed white blood cell count of 1.9 of a few days earlier.

NOT SO BUSY, AFTER ALL

By 10:00 a.m. this morning, my four doctors' appointments had been reduced to one, after Dr. Saltz made a tactical decision, and I made a strategic decision. With free time, Susan, David, and I were able to meet my sister, Katie, for lunch at a local, very good, new Lebanese restaurant. Katie is here from California for a few days and helped Susan put away the contents of our car, which had just arrived from Florida, courtesy of Christophe Clement, who shipped it up along with a group of other vehicles belonging to his stable employees.

A high-end racing stable like Christophe's, comprised of valuable stakes and allowance-class runners, is a complex operation. Not only does a superior trainer have to have perfect pitch for horses that enables him or her to anticipate and react to the individual needs of these high-strung, explosively powerful, sensitive, fragile, and sometimes dangerous athletes, but also the trainer has to deal with a myriad of business, strategic, logistical, legal, regulatory, financial, and veterinary problems. I can think of no other occupation that requires dealing so routinely with the full spectrum of the human condition: owners ranging from kings and queens to brash entrepreneurs; employees ranging from people with options in life who simply love working with horses to people with sad, endless personal problems and no better career choices. What makes all this worthwhile for the trainer and everyone else involved? A good horse. The dream and occasional reality of a top horse keep it all going.

The dream for an American involved in racing is to win the Kentucky Derby. Next Saturday, May 2, is this year's Kentucky Derby, to be run, as always, at Churchill Downs in Louisville. Once again, I won't be running a horse in the Derby. But I am excited about the prospect of running Hot Money that day in a race at Belmont Park on Long Island.

DERBY DREAMS

With the Kentucky Derby only one week away, everyone involved in thoroughbred horse racing is focused on the latest updates about the horses eligible to run—their health, appearance, final workouts, changes in equipment and jockeys, and so on. To run, a horse must be a registered three-year-old thoroughbred that was nominated to the Derby before certain deadlines.

A dramatic subplot has developed around which horses will actually get to run in the Derby. Years ago, for business, social, and sporting reasons, an owner did not enter a horse in the Derby unless he or she thought that it had a reasonable chance to win. There are significant entry fees, and it did not make economic sense to use one of a valuable horse's limited number of career starts in a futile effort in an especially demanding race like the Derby. Moreover, it was considered both a personal disgrace to have your horse exposed as outclassed and a disservice to the horse to ask it to make an all-out effort in a race against superior competition. Few owners wanted to look like fools or as if they were putting their own desire to be part of a Derby ahead of their horses' welfare.

More and more owners have come into racing for whom the glamour of running a horse in the Derby supersedes all other considerations. Some owners now regard running a horse in the Derby as an achievement. As a result, Churchill Downs has had to limit the number of starters in the race to twenty, as determined by the horses' earnings in graded stakes races, the highest quality races. Now that there is a competitive basis for qualifying to run, any remaining stigma attached to running a long shot in the Derby has faded, and most owners who have a sound three-year-old with top-twenty graded stakes earnings feel that their horse has earned the right to run, even if it has little chance to win. Thus, the twenty-horse ceiling has also become a twenty-horse floor. Each year, it seems that at least one talented, but late-developing, horse that would be one of the favorites to win that year's Derby is in danger of being excluded on the basis of qualified earnings. This year, Dunkirk has only made three career starts, but will be one of the betting favorites in the Derby, if he gets to run.

An irony of this twenty-horse syndrome is that it enhances the chances of a long shot winning the Kentucky Derby. With so many horses jostling for position to save ground on the turns, racing luck tends to play more of a role in Kentucky Derbies than in most races. There has always been a lot of uncertainty in picking a Derby winner anyway, partly because the race is over a distance— a mile and a quarter—that few if any of the horses have ever run in their young lives. I haven't checked lately, but at one time it was true that a bet in the same amount on every starter in every Kentucky Derby, he or she would have yielded a profit because of long-shot winners.

KICKING ON

It's hard to live to the end of your life.

—Mrs. William C. Jones,
my Aunt Lucy, recalling from memory a passage in
DOCTOR ZHIVAGO

REAWAKENING TO A NIGHTMARE

On the morning of March 2, 2009, while sitting in the waiting room of a surgeon, I received a telephone call on my cell phone from my regular physician, a gastroenterologist, who informed me that an MRI showed that I had cancerous lesions throughout both lobes of my liver. Within minutes, the surgeon for whom I had been waiting had examined me and told me that he thought that I had a tumor in my colon. I realized that the combination of cancers almost certainly meant advanced metastases and death.

My first thought and action was to call Susan. We had just returned from South America, from which I had arranged tests and doctors' appointments in New York. My next appointment was in a half hour, to see a periodontist. On the one hand, in light of what I had just learned, it seemed inappropriate to tend to such a mundane task as going to a dentist. On the other hand, I had a troublesome dental problem that was going to need attention at some point. What was to be gained from putting it off? I walked through the snow to the periodontist's office. On the way, while observing the cityscapes with a heightened sense of awareness, I focused on the existential questions that had come to mind when I learned that I had what was incurable cancer: what to do with the remainder of my life, how to find meaning in what remained? How should I conduct myself? I felt in great need of guidance, of a framework for thinking. I decided that as soon as I got home, I would start rereading Frankl's *Man's Search for Meaning*.

After a day or so of pondering whether my best course of action would be simply to enter hospice care, I decided to investigate whether treatment would make any sense and began to contact a few friends in the clinical and research science communities who could connect me with oncologists who could help Susan and me decide what to do.

The week after I received the initial news of cancer was consumed by medical consultations, tests, and procedures, contacting family members, and beginning to contact friends, as well as carefully and painstakingly rereading *Man's Search for Meaning*. It seemed as if all of this were happening in a nightmare. Whenever I awoke from sleep, I had the sense that I was reawakening to that nightmare. I never had a conscious moment of denial (unless I am still in denial, which I suppose is possible). I never once thought, "Why me?" After about a week, the nightmare ceased to be a nightmare: It had become my reality.

From the moment that I realized that I had incurable cancer, I sensed that such a diagnosis is inherently isolating and depressing, as dying is the ultimate act of loneliness. Although I accepted that my body had fought and lost its battle with cancer before my mind knew about the cancer, I felt strongly the need to fight against isolation and against depression. Gradually, instinct crystallized into thought: I articulated to myself that there would be no point in enduring the suffering of chemotherapy if I were to allow myself to lose control of my mind to depression, having already lost control of my body to cancer. If I were to allow myself to become depressed, not only would I be incapable of finding joy in at least some part of each remaining day, but also it would make my plight and decline that much more difficult for loved ones, especially family. I concluded that depression was simply a luxury that I could not afford. These are brave words, from a person who is not brave. I do not know how I will handle what's coming.

Having always been a private person, it was unlike me to be writing this blog. Nevertheless, when the notion of writing a blog to keep family and friends informed was suggested to me, I immediately grasped the merits of the idea—not only because it would go far towards solving an inherently difficult communications problem, but also because it could be a potent weapon in the fight I was determined to wage against desolation.

I have been touched by your kindness, the kindness of my friends and family. Because of you, your interactions, your communications, and your prayers and positive thoughts, I am less isolated than I have ever been in my life. I have heard from people from whom I had not heard in over fifty years. Far from being depressed, since the first week after my diagnosis, no day has gone by in which I have not found at least some moments of joy (although I must confess that one day came close). I no longer reawaken to a nightmare, but to the reality of my life, for which there has been, and remains, much to be grateful.

APRIL 28, 2009
TUESDAY 5:18 PM

SPANISH FLU

Ninety years ago, one of my grandmothers—my father's mother—died in the influenza pandemic of 1918. I have long wondered if my father's ability to recover from his contraction of that influenza is related to his longevity. As you would expect, I have always paid attention to potential pandemics, such as the current swine flu.

My impression is that most people assume that, because of advances in knowledge and medicine, something like the 1918 pandemic could not happen today. Although the number of its victims is unknowable—India did not compile statistics, for example—it certainly killed far more people than died in World War I. One estimate that I have seen is that about a third of the world's 1.5 billion population contracted the influenza of 1918, and some fifty million died. Up until several months ago, most people were similarly complacent that we could never suffer the sort of collapse of the banking system and financial markets that the world experienced in the 1930s.

OATMEAL AND OYSTERS

My blood counts were high enough to receive chemotherapy today. Moreover, my red blood cell count was close to normal. As a result, I will not have to be infused with iron sucrose during this chemotherapy cycle. Presumably, the improvement in my anemia is primarily owing to the efficacy of the chemotherapy. But I have also been assiduously incorporating iron-rich foods into my diet, especially oatmeal and shellfish. It was a pleasant surprise to learn that my oncologist at MSKCC did not object to my eating raw food, including clams and oysters. He also does not object to patients drinking wine, and I find that Sauvignon Blanc still tastes like Sauvignon Blanc and still pairs well with oysters. (I've never tried Sauvignon Blanc with oatmeal.)

IN THE SPIRIT OF SCIENTIFIC INQUIRY

Upon finishing the previous post on my blog today, I realized that I had worked myself into a real hunger for oysters. My chemotherapy session had ended on time, I was feeling well, and the lunch hour was approaching. I called a friend who appreciates good food and wine to see if he would be available for an impromptu lunch at the Oyster Bar at Grand Central Terminal. He came to mind not only because he is great company, but also because he is a medical doctor and research scientist who, unfortunately, has lived through the terminal cancer of a spouse and has a great deal to teach me. Moreover, I certainly didn't have to worry that he might be put off that, however surreptitiously, I am hooked to a bottle of 5-FU.

What had started out as a cool, damp morning had turned into a lovely spring day. I thoroughly enjoyed the mile stroll to Grand Central, walking most of the way along a colorful stretch of Second Avenue. After contemplating the wine list at the Oyster Bar, out of respect for my company, I decided in the spirit of scientific inquiry that I should determine if a decent Chablis still tastes to me like Chablis and still works with oysters. I am happy to report that, as of lunchtime today, the answer to both questions is still, "Yes."

MAY 1, 2009

FRIDAY 6:04 PM

KICKING ON

Since I developed a rash on April 20, after receiving a shot of Neulasta for the second time, I have been quite apprehensive about what would happen the next time that I received Neulasta. Given how suppressed my white blood cell count became when I received chemotherapy without Neupogen or its longer-acting version, Neulasta, it is obvious that if I could not tolerate Neupogen or Neulasta, my chemotherapy would have to be curtailed significantly. Neupogen and Neulasta commonly cause bone pain, which I experienced with Neupogen, but not with Neulasta. Neither is supposed to cause a rash, and I had received Neupogen twice and Neulasta once with no skin rash. Dr. Saltz decided to give me Neupogen this time, so that if I got a skin reaction, it would not be exacerbated over an extended period by the longer-acting Neulasta.

I received the injection of Neupogen at 8:50 a.m. today, and when I got home and checked in a mirror about a half-hour later, I saw that I was developing a rash. After several telephone calls with Dr. Saltz's nurse, I was sent to the Urgent Care Center, where I was seen by a doctor who conferred with Dr. Saltz by telephone, and then to another MSKCC facility to be examined by a dermatologist, who conferred with Dr. Saltz by telephone at about 3:30 p.m. Dr. Saltz decided that if my reaction to Neupogen does not worsen significantly, I can be injected with it again tomorrow and each of the following three mornings.

Susan and I got home in time to watch with David the televised Kentucky Oaks, which was won as easily as I have ever seen a horse race won, by the great three-year-old filly, Rachel Alexandra. Had she been entered in tomorrow's Kentucky Derby, I think she would have been favored to win. In one of Susan's and my greatest experiences in racing, we were guests of Bert and Diana Firestone when their filly, Genuine Risk, won the Derby twenty-nine years ago, the second time a filly ever won the Derby. As the field for this year's Derby stands, I think the lightly raced colt, Dunkirk, is the most talented entrant.

We are planning to run Hot Money, one of our three-year-olds, in a six-furlong grass race, the tenth race tomorrow at Belmont Park. It is raining here now, and it is supposed to rain tomorrow, so we have our fingers crossed that the race will not be moved from turf to dirt. He has never run on turf, but his dam was a turf horse, and he moves well on that surface. As it would be over a month before there would be another suitable turf race for him, we plan to run him even if the race comes off the turf. His trainer, Christophe Clement, says Hot Money is training well, looks well, and should be ready to run tomorrow.

HOT DAMN, HOT MONEY

Shortly after returning from the winner's circle, I looked at my Blackberry and saw, to my surprise, that I already had a number of e-mails congratulating us on Hot Money's win today, one of which was entitled, "Hot Damn Hot Money." I hope that the e-mailers had bets on Hot Money, which went off at 8.6-to-1 odds as well. As the track announcer called at the finish of the race, "Hot Money got the money."

Hot Money has a nice story line. Christophe trained Hot Money's dam, Smoldering, for a European owner. Smoldering showed some potential, but slipped and fell on the way to the track one morning and injured herself badly. Saving her would require an expensive operation followed by a long convalescence, and she would be unlikely to race again. Her owner declined to undergo that expense and ordered that she be put down. Christophe then received permission to pay for her surgery himself and took title to her. When I learned about it, I offered to take on the expense, and when my partner, Peter Karches, learned about it, he offered to share the expense with me, and Peter and I became the owners of Smoldering. There are more twists and turns to the story, but Susan and I wound up as the sole owners of Smoldering and bred her to Not for Love, a stallion in which I owned a five percent interest. We named the resultant foal Hot Money.

Hot Money got sick as a foal and was on antibiotics for months. He grew to be a small, but well-made, balanced horse, with one pronounced flaw: He walks with his hind legs far apart for some reason that we cannot determine. For want of a better solution, I had him gelded this winter. He tracks better now, and certainly ran well today in his second lifetime start, the first since he was gelded. He is quick on his feet and can be a bit naughty, unseating his exercise riders in the morning if they are not alert. We hope to win a lot more money and have a lot more fun with him.

SUNDAY SILENCE

I spent the morning at MSKCC getting an injection of Neupogen and a CAT scan. My exercise for the day consisted of walking back and forth from the hospital in light rain, amidst bicyclists who had the run of the city in a sanctioned event. After having lunch and watching reruns of Hot Money's race, which we had recorded on TiVo, I devoted the rest of today to reading by the fireplace in our living room.

TWO CUPS OF COFFEE

As I have written previously, when I learned that I had cancer, I realized immediately from the nature and extensiveness of it that under no circumstances would I live to a ripe old age. Because of my parents' extreme longevity, and because I had long taken great pains with respect to diet, exercise, and medical check-ups, I had always been concerned about the disabilities of old age. In particular, I was concerned about going blind one day. My father suffers from wet macular degeneration, and my mother is legally blind from glaucoma. About ten years ago, I was diagnosed with a complicated case of glaucoma as well, for which I meticulously administer four different prescription eye drops in various combinations each day to lower my intraocular pressure from normal to sub-normal. About seven years ago, I serendipitously discovered that caffeine raises my intraocular pressure. Since then, I have abstained from caffeine, which had always been my drug of choice.

The morning after I learned that I have cancer, I decided that, now that I don't have to worry about outliving my eyesight, I may as well go back to enjoying coffee. So I had a cup with my breakfast, and it was every bit as enjoyable as I had remembered. I did the same thing the next morning. But the third morning, I decided that even though resuming coffee consumption would now be perfectly rational behavior, it seemed like a self-destructive, negative thing for me to do. To date, I have not had another cup of coffee.

MAY 6, 2009
WEDNESDAY 5:00 PM

OPTIMISM

Although most of my family seems to accept the reality of my diagnosis as an incurable, a number of my friends have obviously been disappointed with me for "giving up," that is, for not believing that somehow I can *beat* the cancer that afflicts me or at least survive for a long time. I am deeply touched by their concern.

Some of my friends' difficulties in accepting that my condition is incurable may be owing simply to incomplete information about my diagnosis or incomplete understanding of the resultant prognosis. Difficulty in acceptance by others that I am incurable may stem from a belief that if I "give up," my outcome will be negatively affected. For some, the difficulty may go to the heart of their belief systems, such as beliefs in miracles, miracle cures, divine intervention, and the curative power of prayer and positive thoughts. (I am profoundly grateful for friends' thoughts and prayers.) For some, if I do not believe that I can survive my condition for an extended period of time, it is harder for them to deal with my plight. For some, if I accept my mortality, it makes it more difficult for them to deny this now. Others may simply not understand the distinction between palliative care, which is all that is available to me, and curative care, which for me is off the table.

In spite of my acceptance of my fate, I do not find myself lacking in optimism or positive thoughts. As the recipient since my diagnosis of extraordinary kindness from loved ones, friends, and care-giving professionals, I have never felt closer to my fellow man or more optimistic about the human condition. I see no contradiction between being a realist and being an optimist.

I could describe my current state of mind as that of a realistic optimist. Susan and I recently made another monetary contribution to cancer research, not out of any hope that it will benefit me, but rather out of faith that eventually cancer research will result in significantly lower death rates and longer survival times for others who will be afflicted by cancer.

I enjoyed myself today with friends at lunch and dinner, and I have plans to do the same tomorrow.

MAY 7, 2009
THURSDAY 4:10 PM

SNOWBALLS IN JULY

Today, I learned the results of last Sunday's CAT scan, which confirmed what was expected, given improvements in my blood work, skin color, and energy level and a weight gain of about nine pounds. My tumors have been reduced in size by chemotherapy since my first and only previous CAT scan in early March. Although the tumors have not responded as dramatically ("they are not melting like snowballs in July") as is sometimes the case at this point in chemotherapy, the progress is consistent with continuing the current regimen.

I learned that the dermatologist's initial reaction, when he heard last Friday by telephone that there was a patient who had a skin rash after being injected with Neupogen, was that they would have to discontinue giving me the drug. After examining me, he changed his mind.

MAY 8, 2009
FRIDAY 4:47 PM

GYM RAT

For the first time since we left for South America in early February, I went to the gym today. Although I didn't do much—twenty-five minutes of cardiovascular work and about ten minutes of light weightlifting and sit-ups—I got pleasantly tired. As I had wondered if I would ever be able to go to the gym again, I also had the satisfaction of accomplishment, however trivial.

In early 2002, having determined that I would retire from business at the end of 2008, I decided that I did not want to be physically unable to enjoy fully my retirement years. I went on a diet, which entailed losing about two pounds a month for twenty-four months. I bought new clothes and had my old clothes altered, so that I would have an immediate logistical problem if I gained weight. And I went on an exercise program, which consisted of two hours at the gym every morning (one hour of cardiovascular exercise, one hour with a personal trainer) and about an hour of walking around the city during the day. A number of people, not least Susan, said I got to be too skinny. Vestigial love handles argued otherwise.

MOTHER'S DAY

My right shin and ankle are painful today, so our horses may have to be my proxies for exercise for a while. When the doctors examined my leg last Thursday, they were unsure about the locus and cause of the generalized pain. The pain is more acute now than it was then, so I have a better idea about its nature: Bone is not involved, and it has something to do with circulation. Partly because I note that moist heat seems to help the condition more than cold, my current uninformed opinion is that my chemo-therapy has given me a case of phlebitis, or something of that ilk. My inclination is to wait to go to the hospital tomorrow, when the specialists are all on duty, rather than go to the emergency room today, Mother's Day Sunday. Besides, it is a pretty day, and I'd much rather go out to brunch than sit for hours in Urgent Care.

Our trainer in France, Nicolas Clement, has entered Sacred Music in a race at a track in Strasbourg. The race is on Tuesday, at a distance of 2,350 meters (nearly a mile and a half), for a total purse of 17,000 euros. Nicolas says that Sacred Music has trained well since her recent racing debut, and that the long distance of her race on Tuesday should be no problem for her. Although I don't know the form of horses in France, I think that we have every reason to expect her to give a good account of herself. When we get the entries tomorrow, Nicolas can help us assess the opposition.

We are pointing two of our horses in New York to races at Belmont Park. We have entered Mustang Island for a race on Thursday on the grass, similar to the race that Hot Money won last Saturday. This sprint race on the grass will be the second start of Mustang Island's career, just as last Saturday's race was for his three-year-old stablemate, Hot Money. Two days ago, W.C. Jones worked well over the Saratoga training track, as Christophe prepares him for his scheduled comeback race on May 24. We have been resting and rehabilitating four-year-old W.C. Jones since his last race, which took place about a year ago. He will need one more work before his race. Although it is a lot to ask of a horse to win at a distance of a mile and a quarter (the exacting distance of the Kentucky Derby) after a long layoff, Christophe is celebrated for his ability to have a horse ready to give a peak effort under such circumstances.

Hope springs eternal in the horse business, particularly when one has young horses that have not yet established the limits of their potential. The racing of thoroughbred horses is a sport with origins in England more than three centuries ago. The sport is full of hoary and colorful aphorisms. A favorite of mine is, "No man with an unraced two-year-old ever committed suicide." We have an unraced two-year-old colt, Backslider, training in Kentucky.

MAY 11, 2009

MONDAY 9:40 PM

WHEN TO DRINK CHAMPAGNE

In victory, you deserve Champagne; in defeat, you need it.
　　　　　　　—Napoleon Bonaparte

This morning, at about 10 a.m., I reported to Urgent Care at MSKCC; at about 7 p.m., I left the hospital, having been implanted with a vena cava filter. Cancer has made me susceptible to blood clots. I have blood clots in my lower right leg. The existing blood clots are highly likely to propagate into the upper leg. Because blood clots frequently cause pulmonary embolisms, which kill fifty thousand to two hundred thousand people each year in the United States, the doctors would normally treat me with anticoagulant drugs. Unfortunately, the size of my primary tumor makes anticoagulants inadvisable. MSKCC will monitor the blood clots, which will be an ongoing problem for me, and defer for now making a decision on giving me anticoagulants. The vena cava filter is designed to trap blood clots, to reduce the probability of pulmonary embolisms. It was inserted through my neck into the inferior vena cava, the largest vein in the body, which drains blood from the lower body into the heart.

Tomorrow, Sacred Music will face fifteen other three-year-olds, including eight males, in her race in Strasbourg, the Prix Champagne de Venoge. Whether or not she is victorious, we will raise a glass to her.

MAY 12, 2009
TUESDAY 11:05 PM

COCKTAILS

Sacred Music ran fourth in her race today. MSKCC ran hours late in its chemotherapy suite today. Early reports are that Sacred Music came out of her race in good order. As a result of friends treating us to a bottle of post-therapy Dom Perignon, I am pleased to report that I came out of my chemotherapy session in good order also.

MAY 13, 2009
WEDNESDAY 2:51 PM

LIFELONG TEACHER

A friend of mine passed along to me a quotation, from Henry Brooks Adams, that my friend incorporated in his eulogy for his father, a lifelong teacher: "A teacher affects eternity, he can never tell where his influence stops."

Over an excellent lunch at the Oyster Bar (my red blood count is rising, by the way), this friend, a co-investor and co-director with me in companies that we helped to build, was kind enough to suggest that I had had influence similar to that of a teacher, as a result of mentoring managements of companies and younger colleagues through the years. To my surprise, he said that he himself sometimes thinks about what I might say or do in a given circumstance. Because of the great esteem in which I hold this decent and successful man, I must admit that I was not only flattered at the time, but also find myself pondering what he was saying.

As I think about Adams's quotation, I think it is probably true that all of us influence others, for good and for ill, if only in our personal lives and decisions, far more than we realize. Certainly I have been startled many times over the years when someone has mentioned to me how much he or she has applied in his or her life something I said to them fifteen or twenty years ago—which I invariably have no remembrance of ever saying. (I never had a great memory, even when I was young.) My career did not offer as much opportunity to influence others as does that of a lifelong teacher. Nevertheless, as I think about it, the nature of my career inevitably entailed my affecting a fair number of other people's lives.

From 1966, when I got out of my training class on Wall Street and became assistant to the director of research at Reynolds & Co., I was involved in the recruiting, hiring, firing, and promoting of people throughout my career. As a private investor and venture capitalist, I invested in well over eighty other privately held companies, in many of which I was directly and indirectly influential. Managements and co-investors have to pretend to pay some attention to you when you own voting shares, when you may be willing to invest more, and especially when you sit on the investee company's board of directors and compensation committee.

In my career, I also sat on the boards of publicly held companies and not-for-profit institutions. Most such responsibilities require making some difficult decisions that affect other people's lives. With the best of wills, not all decisions prove to be wise, and certainly many are viewed as negative and hurtful by people directly and indirectly affected. Ironically, many people who are cheered by a decision take the wrong message from it, and many who disagree with a decision, and dislike you for making it, learn from it—perhaps consciously, perhaps not—and improve their lives as a result.

I have been asked what I did best in my career. I think I hit my peak at about age twelve when I trained my dog Colonel, a dalmatian, for obedience trials. Otherwise, I think that I was best at giving people, especially young people, a chance to do things for which they were ostensibly not qualified and that they had not thought to try. Most succeeded at their new opportunities. Sometimes I suppose I mentored them intensively, sometimes not. I never felt comfortable being called a mentor. I felt that I was learning so fast, because I had so much to learn, that I was hardly in a position to regard myself as a mentor.

How does one define eternity? To me, it will last as long as the human race and our planet exist. I do not think that the human race, or our planet, or our solar system, will last forever. But I do accept that there is a secular eternity as long as human beings exist; as an imperfect being, I hope that the good that I have contributed to it outweighs the bad. As I think about it, I attribute most of my failings, other than mistakes in judgment, to cowardice. I am still trying to muster courage. It is never too late for self-improvement.

AFFIRMED

Today, in the ninth race at Belmont Park, we had planned to run one of our horses, Mustang Island. He was the second choice in the race on the morning line. This morning, I got a call from Christophe indicating that Mustang Island came back lame from a routine jog. It seems that he popped a curb (a type of soft tissue injury) on his left hock. If we are lucky, he might be ready to run again in a couple of months.

This Saturday, at Pimlico Race Course in Baltimore, the Preakness Stakes will be run. The Preakness is the second leg of the Triple Crown of horse racing. The Preakness is held two weeks after the Kentucky Derby and three weeks before the Belmont Stakes. No horse since Affirmed in 1978 has won all three races. In my opinion, the only horse that still has the possibility of winning the Triple Crown in 2009, the Derby winner, Mine That Bird, will not accomplish that feat.

In a delightful fluke, Mine That Bird won the Derby as a 50-to-1 long shot, the second highest odds on a winning horse in Derby history. Mine That Bird got pinched back to last of nineteen at the start of the race, giving his jockey, Calvin Borel, the opportunity to gallop him along at the back of the pack, along the rail. The surface along the rail was firmer than the rest of the track (sort of like the strip along the beach where a wave has just receded) as a result of rainfall and track maintenance procedures. With about three-eighths of a mile to the finish line, jockey Borel, in a brilliant ride, was able to begin passing the other horses on the inside, as they drifted off the rail owing to fatigue. Mine That Bird ran without interference on a firmer, faster surface, on the shortest circumference of the course, while the other horses jostled each other for position on a deeper, slower surface.

Even though Calvin Borel is illiterate, he is no fool when it comes to horses. He is switching horses tomorrow to ride the only filly in the Preakness, Rachel Alexandra. I think she is as good a three-year-old filly, at the sort of distance over which the Preakness is contested, as I have ever seen. Tomorrow, for the first time, she will try to beat male horses. I think she can do it.

MAY 15, 2009
FRIDAY 4:27 PM

KEEP WALKING

Since undergoing the procedure last Monday in which I had a vena cava filter inserted through the right side of my neck, I have had increasing pain in my lower back, on the right side, when I walk. My lower right leg also hurts when I walk, because of the blood clots. Being able to walk has never seemed so important to me. I'll try to keep walking.

MAY 17, 2009
SUNDAY 11:27 AM

RUNNING IN STYLE

Although I won my bet, and Rachel Alexandra became the first filly to win the Preakness in eighty-five years, I was wrong about the Derby winner, which finished second. Mine That Bird was no fluke. Apparently, the change in his running style, starting with the Derby, in which he is now taken back at the beginning of his races and urged to run in the last stages of his races, makes him much faster overall from start to finish. His radical change in competitiveness has me thinking about my horses. Before I give up on one of my young horses, I will want to talk with the trainer about the possibility of having the rider take it way back at the start of the race (assuming that its temperament would permit it to be rated).

On Friday, for some reason or reasons that I could not pinpoint (side effects of recent chemotherapy, being hobbled by leg and back pains?) my mood was a little down. On Saturday morning, my mood was lifted immediately upon opening my e-mail and reading several kind messages. Then my sister, Katie, arrived with a present that really reminded me how kind people can be. As a favor, a friend of Katie's had painted from photographs an oil portrait of our two-year-old colt, Backslider. Katie, who lives in California, had flown to Maryland, where the artist, Ann Compton, lives, to bring me the painting. Moreover, Ann Compton had framed it beautifully. The day ended with Susan, Katie, and I watching on television the historic victory of the magnificent Rachel Alexandra. How could I not feel blessed?

MAY 18, 2009
MONDAY 8:51 PM

QUICK QUESTION, LONG ANSWER

This morning, I called MSKCC to find out if it would be okay for me to take Celebrex for my back pain. After eight hours in Urgent Care and in the imaging center, various blood tests, and an MRI scan, I had my answer: "Yes."

For diversion, I was fortunate to have a visitor in the emergency room—a friend since college, who had journeyed down from Connecticut to see me. We had a spirited visit. In fact, out of respect for the other patients and their caretakers, we struggled to subdue our chuckling. We plan to get together again soon, in a more fun environment.

UNCLE WILLY

In the sixth race tomorrow at Belmont Park, to be contested at a mile and an eighth on the turf, we have entered four-year-old W.C. Jones. He last ran eleven months ago, in a similar race, finishing a fast-closing second, beaten a head, after trailing the field at the top of the stretch. Because he has been away from the races for such a long time, recovering from an injury, he may need tomorrow's race to return to his peak form. We plan to be at Belmont to see his return.

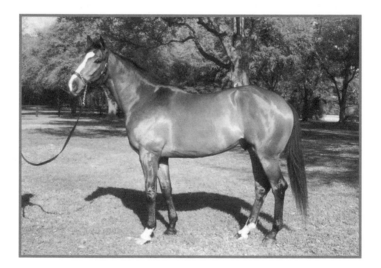

I named W.C. Jones after my late uncle, William C. Jones, whom I called "Uncle Willy." My partner in the horse business, Peter Karches, always said that I was the man who had cost him the most money, because he had never owned horses before he went into partnership with me. Uncle Willy had the same lifelong effect on my pocketbook, by introducing me to thoroughbred horses in 1961. He invited me to the running of the Virginia Gold Cup steeplechase races near the farm where he and Aunt Lucy were raising thoroughbreds. At such country races, the police looked the other way while the bookies operated with their chalkboards in full view. Unfortunately for me, I won the last five of the six races on which I bet, including my bet on Carry Back in the Kentucky Derby, which was run the same day. Two years later, during the summer between my junior and senior years at Princeton, I worked at Saratoga for Uncle Willy's trainer as a hot walker, the lowliest of stable employees.

As Aunt Lucy says, life with Willy was never dull. He was an entrepreneur in the paper industry who always operated with flair, even in later years when he worked as an executive for a company that he did not own. He had grown up riding horses under the eye of his demanding father, and he bred, sold, raced, and fox hunted thoroughbreds in Virginia, and later in New Jersey. He bet on horses at the racetracks in New York and New Jersey, especially on gray horses, which he also preferred to ride. Before breeding thoroughbreds, he had excelled as a breeder of bird dogs, Tennessee walking horses, and prized beef cattle. He always had the most avant-garde equipment for his sports activities, which included at various times tennis (with his own court, of course), golf (at one time, all custom-made woods, no irons), shooting (he used custom-made over-and-under shotguns), and deep-sea fishing (his expeditions were launched from his houseboat, which slept ten). His sporting passions changed without warning, except that he always loved dogs. Most important to me, he gave me my dog, Colonel, for my twelfth birthday.

Uncle Willy was a cool guy. Always a natty dresser, he changed cars constantly. (My father, who also loved to buy a new car, used to say that Willy turned in a car whenever its ashtray filled up.) Uncle Willy drove the first Jaguar I ever saw, a black XK-120 with red leather upholstery and a white convertible top. The Jaguar was followed by a parade of exotic cars, including at least one Aston Martin.

One night while I was a student at Princeton, I was staying at Uncle Willy's farm in New Jersey and borrowed one of his cars so that I could take out a date. I dented the front end of the car and spent the night worrying about his reaction. When I saw him the next morning, he was reading his Sunday paper. When I told him what I had done, he simply asked, without looking up, "Anybody hurt?" Later that morning, when we went to the garage for an outing in another car, his hand-made Mercedes convertible, he walked by the car that I had banged up without so much as glancing at it. Of course, he refused my offer to pay for the repairs.

I don't know how Uncle Willy would have reacted if I had tried to name a horse after him while he was alive. I asked Aunt Lucy for permission to do so at his funeral, and she liked the idea. She saw W.C. Jones a few months after I bought him at a yearling sale and liked his looks and manner. She noted that, unlike his namesake, he seemed easy to train. We have fun discussing his training and racing over the telephone. I'll be calling her at her Virginia Beach home right after the race.

MAY 20, 2009
WEDNESDAY 9:55 PM

THE RETURN OF W.C. JONES

W.C. Jones ran well in his return to the races today. In a field of ten, he finished third, beaten by a nose for second and another neck for first.

CONTRASTING AGENTS

Yesterday, I drank Champagne at Petrossian and Jean Georges. Today, as scheduled, I drank contrast agent at MSKCC.

LONG WEEKEND

When I was healthy and contemplating retirement, I wondered if weekends, even long ones like this Memorial Day weekend, would seem much different from work-week days. If this weekend is representative, they do seem different. The weather is delightful today, and I feel as if I should be outdoors. I should be playing golf, or doing something else that involves physical activity.

For the first time since I got blood clots in my lower right leg and a sometimes debilitating pain in the right side of my lower back, I returned to the gym today. For about an hour, I used five cardiovascular exercise machines and lifted some light weights. Although I subsequently enjoyed the relaxed feeling that exercise followed by a hot shower brings, I am starting to think that I may have exacerbated the pain in my back.

Now that I am in treatment for cancer, which has entailed so many trips to Urgent Care, I feel insecure about the prospect of a three-day weekend. Yesterday, I noticed a rash on my left shin. It was late in the day, and a dermatologist will not be available to examine it until Tuesday. Based on telephone conversations with MSKCC staff, it seems unlikely that the rash is a reaction to chemotherapy: The rash is unilateral, and I last received chemotherapy ten days before I noticed the rash. MSKCC's staff suggested trying to control the rash with cortisone ointment until it can be examined by a dermatologist. As someone with glaucoma, I have to be careful about using corticosteroids, which can raise intraocular pressure.

Nevertheless, I am enjoying the long weekend. Last night, a friend came in from Long Island and took us to dinner. Tonight and tomorrow night, Susan is cooking, and we are having a dinner guest each night. Today, I am reading and listening to music. Perhaps we will go to a museum tomorrow. (I am interested in seeing the remodeled and just-reopened American wing of the Metropolitan Museum.) On Monday, we may go to Belmont Park to see one of America's most important horse races, the Metropolitan Mile. There is never a shortage of things to do in New York City.

MAY 25, 2009
MONDAY 3:54 PM

BACKSLIDER

Part of the appeal for those who own, breed, raise, train, ride, and otherwise work with young thoroughbred horses is watching them develop. They are like flowers. One has to appreciate them each time one sees them, for they will not be the same tomorrow as they are today. They change especially quickly in response to the stress of training.

When I thought about retiring, one of the things to which I was looking forward was being around our young horses. Horsemen often refer to yearlings and two-year-olds as *babies*. This year, as of January 1, the first day of my retirement, Susan and I had only one baby, Backslider. All thoroughbreds in the Northern Hemisphere are deemed to have a January 1 birthday, in order to group them for racing purposes. Backslider's actual birthdate was February 22, 2007.

Backslider's rate of change as a baby has been especially pronounced. This photograph of Backslider was taken on May 25, 2009. By then, he was a more lean, muscular, and mature version of the yearling that I bought in August 2008, at Saratoga.

Even though I have not been able to travel to see my horses since I began treatment for cancer, the decision that I made after I was diagnosed as incurable not to sell my horses has proven to be a good one. Young horses in training are in the ascent—a perfect antidote for someone who is in decline. My family and I enjoy talking about them and look forward not only to their races, but also to each report from their trainers of their progress.

A POSSIBILITY

He who pretends to look on death without fear lies.
All men are afraid of dying, this is the great law
of sentient beings, without which the entire
human species would soon be destroyed.

—Jean-Jacques Rousseau

A YOUNG SIXTY-SIX

At our first meeting on March 12, Dr. Saltz told me candidly that my cancer was incurable, but was worth treating, because I was a "young sixty-six." He based his diagnosis of incurability on a colonoscopy that showed a large tumor in my right colon, an MRI that showed multiple small lesions throughout both lobes of my liver, and manual examination of a hard lymph node in the region of my shoulder and neck. In other words, I had cancer of the colon that had metastasized extensively into both lobes of the liver and appeared to have metastasized further into at least one remote lymph node.

Ordinarily, MSKCC would have ordered a comprehensive, almost full-body (PET) cancer scan as part of the initial fact-gathering. In light of my rapidly deteriorating condition, however, Dr. Saltz believed that there was little time to spare (it was his sense that I was "slipping off the table") and recommended that I commence chemotherapy (specifically, FOLFIRI) immediately. The goal of the chemotherapy was to pull me back from the brink and then give me the possibility of some months, perhaps even a few years, of relatively good quality of life.

Because of the extensiveness of the lesions in my liver, it did not appear to be a candidate for resection. Moreover, if my cancer had, indeed, spread to a remote lymph node, there would be no point to liver surgery, even if my liver were a good surgical candidate. Colon surgery didn't make sense at that time either, as any surgery would require a hiatus of chemotherapy.

Over the weeks, as a consequence of the chemotherapy, I have grown steadily stronger, and the tumors have shrunk somewhat. Eventually I questioned whether surgery was, indeed, out of the question. Some three weeks ago, Dr. Saltz consulted with a senior liver surgeon at MSKCC, Dr. William Jarnagin, who opined that it would be possible to resect my liver with a two-operation procedure. Such a procedure would, of course, make no sense if a PET scan confirmed the presence of cancer in locations remote from the primary tumor in the colon.

Last Friday morning, I underwent the PET scan. Because it seemed so unlikely that the PET scan would not confirm the apparent lymph node involvement, I tried over the last three weeks to deflect any glints of hope. Nevertheless, I had an increasingly hard time sleeping as I waited, first for the scan and then for the results.

This morning, about 8:30 a.m., Susan and I met with Dr. Jarnagin for the first time. He had not yet received the results of the scan, but he described the complex liver and colon resecting that he thought plausible in my case. The risks would be what you would expect: Bad things could happen. The goal would be some years of life, with the possibility of cure. According to Dr. Jarnagin, approximately half of those who underwent such a procedure might live for five years, and perhaps twenty percent might live for ten years, after which recurrence of the cancer would be relatively unlikely. (There is little data. It is my understanding that livers like mine were not resected five years ago.)

Susan and I proceeded to an appointment with Dr. Saltz. After not too long a wait, he met with us. He had received the results of the PET scan and had discussed them with the radiologist. There is no sign of cancer in parts of my body remote from the primary tumor.

The first operation is scheduled for June 15, to allow my body time to recover from the chemotherapy. Another time, I will go into more detail about the planned procedure. In the meantime, I thank you, my family and friends, for your support. If I have had any pretense of courage since the diagnosis of my cancer, it is a reflection of your kindness, your warmth, your prayers, your positive thoughts.

COLLEGE REUNION

Princeton University's annual reunions start today. Among colleges and universities, Princeton has the reputation of being one of the most successful at getting alumni to come back for reunions and to contribute annually to the endowment.

I confess that I have not done my part. I can recall attending only one and part of another reunion, and I have added only modestly to Princeton's bulging coffers. I lived in the present and the future and was busy with my family, work, and social life: Reunions could always be deferred for another year, or even until another big reunion year, like the fifth or the twenty-fifth. Although I had great respect for Princeton and felt fortunate and proud to be an alumnus, I was never able to convince myself that the university with the largest endowment per student was the one that most needed my pittances. Even though research universities have been my philanthropic priority, I gave relatively little money to Princeton, partly because it does not specialize in cancer research.

By the beginning of this year, my perspective on giving to Princeton and on reunions began to change. I had begun studying a relatively new program for contributing to university endowments, in which one could establish a charitable remainder trust that would be commingled with the endowment and participate in its returns. Such a trust would be attractive to someone who seemed likely to live for many more years. As a retiree, I would have plenty of time to attend class reunions, especially big ones like my class's forty-fifth reunion this year. Moreover, Susan and I had accompanied a group of my Princeton classmates and their wives to East Africa two years ago and had enjoyed their company immensely. We made hotel reservations, and I signed up for the reunion.

By the beginning of March, when I returned from South America and was diagnosed with cancer, I had already begun showing signs of cancer cachexia, which is "characterized by weakness, poor appetite, alterations in metabolism, and wasting of muscle and other tissues," to quote Sherwin Nuland from *How We Die*. I was in no shape for reunions, and my condition would have detracted from the high spirits of the occasion. I canceled our reservations.

It turns out that I feel good enough today to have attended the reunion. Chemotherapy's temporary suppression of the cancer has reversed most of the signs of cachexia, and the debilitating effects of the chemotherapy itself have faded in the fortnight since it was last administered to me. Several classmates have told me that they will lift a glass in my honor, and I will reciprocate with the one glass a day of wine that I am permitted until my surgery.

MAY 29, 2009
FRIDAY 7:35 PM

SPRING VACATION

Except for next Tuesday, when I have a full day of appointments at MSKCC, I feel that I have a bit of a spring vacation until my surgery on June 15. For the sixth day in a row, I went to the gym. Although my workouts are light, I am now spending nearly two hours a day in the gym. My objective is to be in the best possible physical condition for the surgery.

I've got a lot of work to do. Having gained eleven or twelve pounds since starting chemotherapy in early March, I'm having trouble fitting into my trousers. Getting in a little better shape may not make any actual difference in the outcome of the surgery. Nevertheless, I think that taking action to try to help one's cause always improves one's state of mind. Because I am getting a little stronger every day, I am enjoying my sessions in the gym as an activity that gives me a sense of progress during my spring vacation.

MAY 30, 2009
SATURDAY 1:07 PM

RACHEL ALEXANDRA

A few people have asked me about the decision of the principal owner of the great filly Rachel Alexandra not to run her in the Belmont Stakes next Saturday. Although I was looking forward to seeing her in the race, I am relieved that she is not running. Modern thoroughbreds need five or six weeks to recover fully from a peak effort such as the one she gave in the Preakness, which she won only fifteen days after winning the Kentucky Oaks in a brilliant performance. If she were mine, I would have skipped the Preakness and run her in the Belmont Stakes instead. My program would have given her five weeks and a day to recover from her effort in the Oaks. Given that she did run in the Oaks and the Preakness, I would not run her in the Belmont Stakes.

Few colts, let alone fillies, run in all three of the Triple Crown races—the Derby, Preakness, and Belmont Stakes. Those that do run in all three are seldom the same afterwards. Although the Derby, Preakness, and Belmont Stakes are great races, they were not created as part of a series; a sportswriter in the 1930s, Charlie Hatton, linked them as the *Triple Crown*. Although I admire the horses that do manage to run in all three of those races and remain sound, I do not think that I would ask a horse to take the risks associated with such a demanding feat.

TESTS OF PATIENCE

Owning and breeding racehorses is a test of patience that usually ends in disappointment, occasionally in exhilaration, and sometimes first one and then the other.

There is potentially a race at Belmont Park for Hot Money next Saturday, June 6. As June 6 is Belmont Stakes Day, the biggest day of racing each year at Belmont Park, owners are anxious to showcase their horses by running them that day. We won't know for a few days whether Hot Money will get a chance to run next Saturday. Even if his race does get included in the program, and he draws in the race, we will then have to worry about the weather: If it rains, the race might be taken off the turf. If the race were taken off the turf, we would have to worry about the condition of the main track and make a decision to scratch or run. Even if he does get a chance to run next Saturday, we will still have to worry about racing luck: How will he break from the gate? Will he be forced wide or get boxed in? And so on.

Why do owners—kings, commoners, and me—subject themselves to such aggravation? In a winning race, there is a moment when you realize that your horse is going to win. The anxiety that precedes that moment builds tension, the release of which is integral to the rush of winning. A horse can become an owner's avatar. The triumphant feeling generated by a long-awaited victory by a beautiful animal, the owner's competitiveness incarnate, is singular.

CHARLIE OF ARABIA

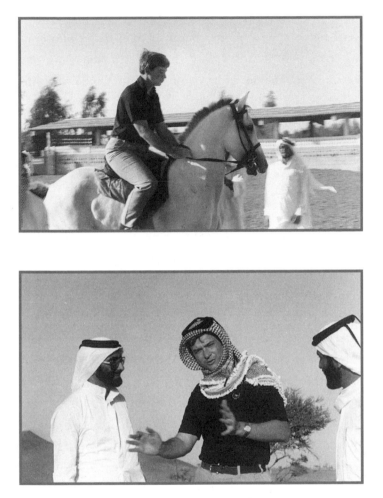

Mary and Martin Corrado, American friends living in Abu Dhabi, just sent me the two photographs above, which they entitled "Charlie of Arabia." I wasn't with T.E. Lawrence; these snapshots were taken on a business trip about thirty years ago, back when I still had hair. The photo at the top was taken at a stable in Al

Ain, an oasis at the nexus of Abu Dhabi and Oman, just south of Dubai. The Arabian horse had a hard mouth, and I remember him well. I think that the other photo was also taken in Al Ain. Wherever we were, it looks as if I were horse-trading.

As I have mentioned, I tended to live in the present and the future. I never kept a photo album or took many photographs. Seeing these two photos, I wish that I had taken more photos. They bring back memories of an adventurous time in my life.

JUNE 2, 2009
TUESDAY 7:47 PM

A POSSIBILITY

Today, three months after my cancer diagnosis, I had four appointments at MSKCC in preparation for the surgery scheduled on June 15. The first appointment was an ultrasonic reexamination of my right leg to determine the status of the blood clots that led to the insertion of a filter in my vena cava (the blood clots have not migrated and have diminished significantly in size). The second was a follow-up meeting with Dr. William Jarnagin, the liver surgeon. The third was an initial meeting with Dr. Martin Weiser, the surgeon who will resect my colon. The fourth was for pre-surgical testing. On June 9, I will meet with the physician who will clear me (presumably) for the surgery and be available to deal with any issues that might arise during my hospital stay that would be outside the responsibilities of the two surgeons.

A lot has happened to me in the last three months. As I am tired tonight, I will post tomorrow what I learned today about my upcoming surgery.

RACING TO SURGERY

Their surgical plan for me encompasses three major surgeries and one incidental surgery. On June 15, two surgeons will operate on me. In the first surgery, the colon surgeon, Dr. Weiser, will remove my right colon in an open-incision procedure that will take about an hour. In the second surgery, the liver surgeon, Dr. Jarnagin, will remove the two cancerous lesions that have been identified on the left lobe of my liver, one of which is in a difficult location from a surgical point of view, plus any other cancers that he may find on the left lobe. In the incidental surgery, Dr. Jarnagin will remove my gallbladder because it's "in the way" of the liver surgery. Finally, Dr. Jarnagin will tie off a major source of blood supply to the right lobe of my liver, in order to make the right lobe atrophy, which will stimulate the regeneration of the left lobe.

The third major surgery will be performed by Dr. Jarnagin six to eight weeks or more later. In it, he will remove entirely the right lobe of my liver. The idea is to cut out all discernible cancer in my body and leave me with a large enough liver to avoid liver failure. The liver will then regenerate. Although I do not yet know if I am to receive chemotherapy between surgeries, chemotherapy will resume sometime after the third major surgery, as the odds of recurrence of the cancer are high. (It is my understanding that, when cancer is said to recur, it actually was never eliminated from the body.)

The surgeons' expectations for June 15 are that, not including preparation time, the colon surgery will take about an hour, and the liver surgery will take at least three hours. Afterward, I will stay overnight in the recovery room and then be moved into a hospital room. I will be gotten up to walk starting the day after surgery. A couple of days after surgery, I can begin to take liquids and, after a few more days, solids. I will have a device that will, within limits, enable me to administer my own pain medication. Six days to two weeks after surgery, I will be released from the hospital. It will take six to eight weeks or more for me to regain strength and full powers of concentration. During the initial stages of my recovery, I will be under instructions not to lift anything heavier than a newspaper or a glass of water. Eventually, following recovery from the second surgery, I can look forward to resuming a normal diet. Dr. Jarnagin told me not to give away my wine collection.

As I learned only eight days ago that surgery was a possibility for me, I am still adjusting my thinking to this new course of action. It seems to me that, even though the June 15th surgery is an important event fraught with risk, it is intended to be but a part of a multi-stage process. If I am lucky, I will have to stay the course in a multi-year marathon, while being thankful for life each step of the way.

JUNE 4, 2009

THURSDAY 4:16 PM

THE MAKING OF A SCULPTOR

This morning, I had a delightful visit from a sculptor, Julio Sanchez de Alba. In 2005, I bought one of the thirty-five castings of his first piece, Tantor, a bronze of an elephant that must weigh several hundred pounds. Julio is a mechanical engineer from Bolivia who got laid off from his job in 2000. He proceeded to teach himself sculpture and metallurgy in the course of creating Tantor. Since then, he has pursued a career as a full-time sculptor. Judging from the public commissions that he has received, it would appear that he is a commercial success. Having seen his enthusiasm for his work and his ability to create bronzes of all sizes starting from tinfoil models, I think that if I have ever met anyone who has in midlife found a field for which he has perfect pitch, it is Julio Sanchez de Alba.

PAY DIRT

Dunkirk was my bet to win this year's Belmont Stakes. He finished second to long-shot Summer Bird. That result was typical of my bets today—I cashed only one ticket.

As poor as my handicapping was, it could not ruin my day. Susan, David, and I had fun, as did our guests. We enjoyed excellent sight lines from our new box. (We recently swapped the box on the front row at Belmont Park that we had rented for many years for one out of the sun, which I am instructed to avoid while on chemotherapy.) The weather was perfect, the racing was first-rate, and I saw many friends and acquaintances in the racing world that I had not seen since last August at Saratoga. The hard part of my day was informing some of those friends about my condition.

Susan, David, and I hope to return to Belmont Park this Wednesday to watch Hot Money run. He is entered in a seven-furlong turf race, and he is the probable favorite to win. Thunderstorms are expected to pass through our area on Tuesday and Wednesday. If the race is taken off the turf, we will have to examine the condition of the main track and decide whether to run him or scratch him. If we run him on dirt and he performs poorly, we will have taken a race out of him at a time when he is training exceptionally well. He trains primarily on dirt tracks. At some point, it would behoove us to find out if he can run as well on dirt as he does on turf. In this country, there are many more races on dirt than there are on grass.

DADDY BILL

We grandchildren affectionately called my maternal grandfather "Daddy Bill." William C. Kiker died on April 11, 1953, when I was barely ten years old, and I never got over it. He was only sixty-seven years old when a heart attack killed him. That was not considered to be an old age, even then, especially for someone who was not overweight, did not drink or smoke, whose father had lived past eighty, and whose mother was still living.

Daddy Bill's mother was born in 1864 toward the end of the War Between the States. I recall visiting her on the North Carolina farm where Daddy Bill grew up, which had been an English land grant to his ancestors. Daddy Bill's father specialized in growing seed corn, and he was able to put all five of his children through four years of college.

Daddy Bill has always represented for me a gold standard for how a man should live his life. A portrait of him hangs in my home office, and I have studied it often when faced with decisions with ethical considerations. He graduated in 1909 with a degree in civil engineering, and won the honors prize in mathematics, from Trinity College in Durham, North Carolina (which later changed its name to Duke University). His bride-to-be, my grandmother Blannie Emmie Berry, graduated magna cum laude in the same class, with honors in Latin and English. He and one of his brothers, Paul, played on what is now recognized as Duke's first basketball team. Daddy Bill received a Duke varsity letter "D" retroactively for the years 1907, 1908, and 1909. (When my son heard about the basketball legacy, he asked me, "Dad, whatever happened to the athletic genes in the family?")

Early in Daddy Bill's career, in the time of Pancho Villa, he worked as an engineer in Mexico on railway construction. He worked also on the construction of the Florida East Coast Railway across the Florida Keys. In 1918, he moved to the little town of Reidsville, North Carolina. In the early 1920s, he organized his own construction company, subsequently building roads, bridges, and airports. He also was an entrepreneur and an investor. Among other things, he and two partners operated a lucrative ferry across the Pee Dee River, and he owned several farms and a small chain of furniture stores. During World War II, he constructed military facilities.

During World War II, my father closed his architecture practice in Jacksonville, Florida, to work for the U.S. Army Corps of Engineers. Subsequently, he and my mother moved to Reidsville, where he joined Daddy Bill's construction firm to build military facilities. I was born on February 6, 1943, in Greensboro, North Carolina, which had the hospital maternity ward nearest to Reidsville.

Daddy Bill was a true philanthropist. Not only did he give generously to institutions, such as Duke University, but also he gave directly to needy individuals, including food handouts to hobos at the kitchen door. Always a sports fan, when the local high school needed a stadium, he built and donated it. He helped to start the local hospital, which he subsequently served as chairman. A civic leader and pillar of the local Methodist church, he served as a member of the City Council and the County Board of Education. He was a charter member of the local Rotary Club, and he took a year off from his business interests to serve as regional governor of his Rotary Club District.

What I remember personally was Daddy Bill's charismatic warmth, humor, integrity, wisdom, and kindness. He always wanted to be present for the birthdays of my older brother, Bill, and me. We would travel out of Jacksonville's July heat to North Carolina for my brother's birthday, and Daddy Bill and my grandmother would come down to Florida in February for mine. There were a lot of kids in our neighborhood in Jacksonville, and they would trail after him as he went on his daily walks. He delighted in children, and he would enthrall us with his observations on our wooded Florida neighborhood's fauna and flora and on the constellations in the sky. When he came to Florida in warmer months, he loved to swim in the ocean—never in pools, which he considered unsanitary. He would swim far offshore, beyond the waves. In North Carolina, we would sit on his large screened porch at night and watch the fireflies, which were much more abundant then than now.

Daddy Bill and my grandmother took me and a woman, Mary, who had worked for them for many years as a domestic, on a visit to New York City. I had never been to New York, and, as Daddy Bill said, "Mary had never a chance to go anywhere." When we went to have dinner at the restaurant at the top of the Empire State Building, the restaurant would not seat us because Mary was black. We went elsewhere for dinner.

I remember once in North Carolina being in a car that Daddy Bill was driving down a dirt road through woods at night. Suddenly, he stopped the car and got out. He had seen a figure beside the road. He helped the drunken black man into the car with us and drove him to the hospital.

Nothing has ever made me prouder than walking along the sidewalks of Reidsville's main commercial street with Daddy Bill on a Saturday, when the farmers came to town. It seemed that everyone—bankers, merchants, farmers—knew him and came out to greet him, with obvious affection and respect; and he always introduced them to his grandson. I never had more fun at sporting events than I had going with him to the class D, minor league games of the Reidsville Luckies. (Reidsville's major industry was the American Tobacco Company, manufacturer of Lucky Strike Cigarettes.)

I have always known that I was lucky to have had such a role model. But I have never felt that I could live up to Daddy Bill's standards. Two days after he died, the local newspaper, *The Reidsville Review*, included in an editorial on his life the following words: "He developed notable ability in managing and directing business affairs, achieving conspicuous success. Much more important was his unselfishness, for as his business enterprises increased in importance and magnitude, he committed responsibility and authority to young men who under his wise guidance learned how to manage and direct important ventures. He was wise in counsel, loyal in friendship and exemplified the qualities of good citizenship." The editorial concluded: "Although curtailed in his customary activity for the past two years or more, his affability and cheerfulness did not diminish. A good, useful citizen has departed from us but the memory of his kindness, generosity and countless deeds of helpfulness will ever be cherished."

JUNE 8, 2009
MONDAY 8:54 PM

TOM POLLOCK

This morning, I received a call that Tom Pollock had died in the hospital. Tom and I were schoolmates and friends from the fourth grade, when his family moved to our school district, through the tenth grade, after which I went away to school. I always thought that Tom was the smartest of our group of friends, and he became a Russian language and cultural expert for the National Security Agency. Tom and I had lost track of each other for over half a century, until we were reconnected by another childhood classmate, George Martin, when George learned that we both had colon cancer.

Tom's surgery in the middle of April seemed to be a complete success. He was released from the hospital and went home. Shortly thereafter, he developed the complications that presumably led to his demise. According to the account that I got, he slipped away peacefully, in no apparent physical pain. I hope that he was at peace in his final days and hours.

JUNE 9, 2009
TUESDAY 8:09 AM

HOW WE DIE

Since being diagnosed with cancer, I have read surprisingly little. The only two books that I have reread are Frankl's *Man's Search for Meaning* and Nuland's *How We Die*. I have always thought that both of these books should be on everyone's short list of essential reading.

Nuland gives what I take to be a clinically realistic, but nonetheless sensitive, account of dying. He makes it clear that "we rarely go gentle into that good night." The practical value of reading his book, for me, is to have a better idea of what one might expect when dying and perhaps somewhat ease the process thereby. As Nuland points out, "In his essay 'Use Makes Perfect,' Michael de Montaigne suggests that a lifelong acquaintance with the ways of death will soften one's final hours." In the epilogue to *How We Die*, Nuland writes about his personal expectations for dying:

> The only certainty I have about my own death is another of those wishes we all have in common: I want it to be without suffering. There are those who wish to die quickly, perhaps with instantaneous suddenness; there are those who wish to die at the end of a brief, anguish-free illness, surrounded by the people and the things they love. I am one of the latter, and I suspect I am in the majority.
>
> What I hope, unfortunately, is not what I expect. I have seen too much of death to ignore the overwhelming odds that it will not occur as I wish it. Like most people, I will

probably suffer with the physical and emotional distress that accompany many mortal illnesses, and like most people I will probably compound the pained uncertainty of my last months by the further agony of indecision—to continue or to give in, to be treated aggressively or to be comforted, to struggle for the possibility of more time or to call it a day and a life—these are the two sides of the mirror into which we look when afflicted by those illnesses that have the power to kill. The side in which we choose to see ourselves reflected during the last days should reveal an image that is tranquil in its decision, but even that is not to be counted on.

JUNE 10, 2009

WEDNESDAY 9:54 PM

A GOOD TURN OF FOOT

In the seventh race at Belmont Park today, a seven-furlong race on the turf, Hot Money won his second consecutive start. Although he was favored to win, the race was exciting for us, as he came from eighth place at the top of the stretch to win by a head. The turf course was listed as *yielding* today and *good* on the day of his previous win, which means that he has yet to run on a *firm* turf course. Because Hot Money has the speed of a sprinter combined with "a good turn of foot," that is, the ability to accelerate quickly, his trainer thinks that Hot Money will be at his best on a firm turf course.

As horses compete above the level of claiming races, they are moved into progressively more difficult competition, with higher purses. Thus, even though Hot Money has closed fast to win both of his starts on the grass in impressive fashion, he will have to keep improving to keep winning. If he does, indeed, perform better on firmer turf courses, he may well be able to keep moving up the competitive ladder. Meanwhile, he is doing for us what a promising young horse does for its owners: He is giving us cause to look forward to his races.

Sacred Music is slated to run at Longchamp in Paris on June 16, and there is a race for W.C. Jones at Belmont Park on June 20. If two-year-old Backslider continues to progress, he may be ready to make his debut within the next few weeks. Mustang Island is recovering well from a hock injury and should be ready to make the second start of his career within the next couple of months. As I approach my first surgery, I am looking forward to all of these races.

JUNE 11, 2009
THURSDAY 8:05 PM

MY SURGICAL PROCEDURE

Although the surgeons have the final say, it appears that I have been cleared for my surgical procedure by the internist, Dr. Adrienne Vincenzino, who was assigned to review my test results and examine me. She examined me on Tuesday and detected a heart murmur as well as a resting pulse of about forty-five beats per minute. As I have never had heart problems, she thought the murmur was probably just a result of my anemic state. Nevertheless, she ordered an echocardiogram. On Wednesday about 12:30 p.m., I got the news that the cardiologist who interpreted the cardiogram found no abnormality. In recent years, my resting pulse was about 50. Over the last year or so, it had gone up to about 60. No one has explained to me why it has dropped to about 45 currently—I am on no new medications.

After fasting and otherwise preparing myself as instructed on Sunday, I will report to MSKCC for surgery on Monday. Following the surgery, Susan will post updates on my condition on this blog.

LAST CALL

In two separate phone calls, I received final instructions today on preparing for my surgery. After midnight Saturday, I am to ingest no solid foods. On Sunday, I am to drink plenty of clear liquids, in conjunction with the same sort of bowel prep that one has before a colonoscopy. After midnight Sunday, I am to eat or drink nothing, though I may use prescription eye drops and brush my teeth. At 12:15 p.m., I am to report to MSKCC. At 2:15 p.m., the surgery is to begin.

Late in the day, I called MSKCC to report that I am experiencing mild pain in my lower right abdomen. Pain in that same area had caused me to seek the medical attention that led to the diagnosis of cancer. The two surgeons, Drs. Weiser and Jarnagin, who are to operate on me Monday were to be notified by e-mail of my discomfort. If the problem worsens, I am instructed to go to the Urgent Care Center over the weekend.

For days after my operation, I will not be permitted solid food. To prepare myself, I have been eating well all week. In addition to patronizing good neighborhood restaurants, I have had the pleasure this week of eating at Adour, Le Bernardin, Milos, Le Cirque, Gotham Bar and Grill, and the River Club. Susan and I are being taken to lunch tomorrow by some friends to L'Absinthe, a first-rate brasserie; and we are taking some friends to dinner at La Grenouille, the last of the classical, formal French restaurants in New York City. It is too bad that I have had to restrict myself to less than a glass of wine per day.

EQUANIMITY

It seems to me that it would be desirable to arrive for a major surgery in as much of a state of equanimity as possible. Although I have had five abdominal surgeries, they were all routine. I cannot recall trying to prepare myself mentally for any of them.

Today was calm. I slept late. We had lunch and dinner at superb restaurants with friends. (A special treat was a ride to and from dinner in a Maybach.) I read Saturday newspapers, and I went to the gym for two hours. In telephone conversations with Nicolas Clement in Chantilly and Christophe in New York, I learned that on Thursday there is a race at Longchamp for Sacred Music and a race at Belmont for W.C. Jones. I am about to watch a favorite old movie, *The Spy Who Came in from the Cold*.

I was heartened by conversations with my children and my parents, as well as with friends. The letters and e-mails that arrived during the day were also balm for my spirits.

DOGS IN THE PARK

In order to maximize my energy for tomorrow's surgery, a nurse at MSKCC advised me to go for a walk today for exercise, rather than go to the gym. Although a session at the gym would have been more relaxing, it was a beautiful day for a walk—sunny, just warm enough to wear a short-sleeved shirt. David and I walked for a little over half an hour, to the 51st Street overpass, and down alongside a dog park and the East River. The dogs in the park were in high spirits.

I feel as if I am returning from vacation to school or to work, but with an ominous overtone. Tomorrow, I will surrender myself to a procedure in which I will give up control and consciousness and emerge in a fragile, vulnerable, painful condition. But what wouldn't I have given a few weeks ago to have the possibility of longer survival or even a cure—the possibility for which these surgeries are a prerequisite?

As I do chores and otherwise make final preparations for tomorrow, I am mindful of, and feel truly blessed by, the support of family and friends. Thank you.

IN THE
RECOVERY ROOM

It's not that I'm afraid to die,
I just don't want to be there when it happens.

—Woody Allen

LATE START

POSTED BY SUSAN:

Charlie did not leave the pre-surgical area for the surgical suite until 4:00 p.m., and I understand there may be as much as an hour of preparatory work before the surgery begins. Dr. Jarnagin will operate first. It will be some time before we have any news at all.

Thank you for the many, many good wishes and positive thoughts sent our way in the last day or two. We really appreciate everyone's caring and concern.

NO SURPRISES

POSTED BY SUSAN:

The surgery will be completed in about twenty minutes; Charlie is just being closed up right now. Both doctors reported that everything went exactly as planned. There were only the two tumors in the left lobe of the liver, and Dr. Jarnagin was able to remove both of them. The one in the awkward position had shrunk considerably with the chemotherapy, and he was able to remove it with margins with which he is comfortable. There were no complications in the removal of the right colon. David and I are happy with the reports.

Charlie will be in recovery overnight. He will be sleeping and will not be aware of a five-minute visit from David and me, which we will be allowed to make in an hour and a half. We will go home after that for the night and return in the morning, when Charlie will be expected to be in his room.

IN THE RECOVERY ROOM

POSTED BY SUSAN:

David and I paid a brief visit in the recovery room. Charlie recognized us and appeared pleased with my report of the success of the surgery, but he was groggy. We are home now and will be over again in the morning, at which time he should be in his room, number 1636.

JUNE 16, 2009
TUESDAY 8:38 AM

THE MORNING AFTER

POSTED BY SUSAN:

The phone rang in my bedroom this morning at 6:20 a.m., and it was Charlie asking me to bring over his glasses right away. Clearly he is making progress. He was in a chair sitting up when I arrived. He is very tired—he's had little sleep for two days—but he's doing extremely well, according to the medical staff and by my observation. He will be up and walking a couple of times today, and Dr. Jarnagin's associate said that Charlie would feel like a real person tomorrow.

COMPETITIVE SPIRIT

POSTED BY SUSAN:

Charlie is stronger and more alert this morning. He had a fever during the night, but the private duty nurse had him get up and walk the hall, and that brought it down. Walking struck me as an odd antidote to fever, but Dr. Jarnagin's nurse practictioner confirmed this morning that the private duty nurse had done just the right thing.

As any of you who have had surgery will know, breathing exercises and walking are an important part of avoiding pneumonia, which is one of the most common complications. Charlie has a device with gradations that he needs to suck at least ten times an hour. The harder he draws a breath, the higher the plastic marker rises. He is not only interested in his own progress—his ability to raise the marker ever higher—but in how high other patients have raised their markers. Doing laps in the hallway, he wants to know the standard for patient laps. As long as we can restrain him so he does not get too fatigued, I am sure this competitive spirit will serve him well in his recovery.

Dr. Weiser indicated that last night and today tend to be the peak pain days. Charlie controls a morphine drip with a button. When he's not moving, he does not have pain. When he's going to be getting out of bed, walking, or doing breathing exercises, he pushes the button.

Overall, Charlie is doing well in his recovery. Perhaps tomorrow he will want to dictate something himself to post.

PROGRESS

POSTED BY SUSAN:

Charlie's color looks better this morning, and the medical staff says he's making good progress. He had a fever again last night, but it's down this morning.

Today's nurse, Susan Hertz, on hearing that Charlie did nineteen laps of the hallway yesterday (fourteen equals a mile), suggested that a balance of activity and rest is important. He seems to have taken her advice to heart, as he promptly got back in bed and went to sleep.

News from the horse racing front has not been good: Sacred Music finished fifth in her race at Longchamp this morning, and it's raining hard here, so W.C. Jones's race at Belmont Park this afternoon will probably be "off the turf." We imagine that our trainer will decide to scratch him. We did get a report on Monday of a good workout for our two-year-old colt, Backslider, at Keeneland.

GOOD NEWS

POSTED BY SUSAN:

Dr. Jarnagin was just in the room and said that Charlie is doing so well that Dr. Jarnagin thinks that Charlie may be able to go home this weekend. He seemed noticeably better to me today, although he is still fatigued and sleeps a bit after any exertion, and I'm pleased that my observation has been medically confirmed.

Thank you again to everyone who has e-mailed or left voice messages with well wishes. I pass on all of them to Charlie.

FIRST BREAKFAST POSTPONED

POSTED BY SUSAN:

When I arrived this morning, Charlie was happy to report that the doctors had said he could order his first solid food since last Saturday. After much deliberation, he ordered scrambled eggs, oatmeal, and apple juice. In an incident rather typical of hospital life, about ten minutes later, a nurse arrived saying that at 10:30 a.m. he would be taken for a CAT scan (part of a research study of Dr. Jarnagin's in which he agreed to participate). You cannot eat before an abdominal CAT scan, and 10:30 a.m. is the cutoff for breakfast orders, so the nurse suggested reheating the breakfast delivered at 9:15 a.m. when he returns sometime after 11:00 a.m. That sounds like a delightful first meal.

I would like to report that Charlie is accepting all such incidents with good humor, but that is not the case. Still, I think his prickliness is just another sign of his continuing improvement.

HALLWAY LAPS

POSTED BY SUSAN:

We are still hoping for a hospital discharge tomorrow, but a decision will not be made until perhaps this evening. Charlie feels stronger all the time, but he still has a considerable amount of pain around the incision. He says that yesterday's hallway laps totaled 25.5, equal to 1.8 miles. I lost count about midday.

BACK AT THE POST

POSTED BY CHARLIE:

Less than six days after being admitted to the hospital, I have just arrived home. Prior to the surgery, the surgeons had estimated that I would be released from the hospital sometime between six and fourteen days after the surgery. I was interested to learn that the appendectomy that had been performed on me about fifty years ago had pushed my colon upwards and to the right. The pain on the right side of my abdomen that caused me to seek medical attention was owing to this unusual location of the colon (colon cancer does not usually present as a painful condition). I do not know whether or not the displacement of the colon had anything to do with the cancer or the failure of colonoscopies to detect the cancer.

JUNE 24, 2009

WEDNESDAY 11:54 AM

DAY BY DAY

POSTED BY SUSAN:

Charlie wanted to wait for some sort of major news before posting again, but I thought some of you might be wanting an update before then. I would describe his progress as incremental, a little bit day-by-day, but with a startling contrast as compared with a week ago. This morning he ate and enjoyed a full breakfast—oatmeal and juice before I got up, and eggs, bacon, and toast with me. A week ago, he ate a little JELL-O and some fruit ice. Today he spent an hour or two with the newspapers. He didn't touch the newspapers we delivered the whole time he was in the hospital.

He's walking two or three times a day over to the East River and back. One or two laps at a time around the hospital hallway was last Wednesday's exercise.

A BUMP IN THE ROAD

POSTED BY SUSAN:

I had intended to entitle this post "Setback," but after hearing this morning's medical report, I decided to change the title to the above. I am hopeful that this new title better fits the situation.

Last night, after two consultations with the doctor on call, Charlie and I returned to the Urgent Care Center. The area around the incision had become redder, and the redness was spreading. At 2:30 a.m., after blood tests, cultures, and opening of several stitches to clean the wound, a bed was located in the completely full, in-patient part of the hospital, and Charlie was readmitted. He is on intravenous antibiotics.

This morning, the doctor said the infection is already better, meaning that it is responding to the antibiotic. The plan is to keep Charlie in the hospital one more night. After he goes home, a visiting nurse will change the dressing daily so that the area can be kept open for continued drainage. Dr. Jarnagin's Fellow (MSKCC term for Resident) says this sort of infection happens in about ten percent of cases, and he appears optimistic about the outcome. I will post again later today with an update.

"NOT EVEN A SPEED BUMP"

POSTED BY SUSAN:

Having no knowledge of the blog or my post, Dr. Jarnagin just told me that the infection is "not even a speed bump." He said it would have no effect on anything from here on out, noting that the infected place is small and already responding well to the antibiotics. He said that the area would be completely healed in about two weeks.

JUNE 27, 2009
SATURDAY 12:12 PM

RETURNING HOME

POSTED BY SUSAN:

Charlie will be returning home about the middle of the afternoon. We may walk from the hospital, which is probably seven-tenths of a mile away. Should they be needed, there are several benches along the route. Yesterday he did forty-eight laps of the hospital hallway, but only four at a time, so the challenge will be the sustained effort, not the total distance.

The infection is healing well, but the open portion of the incision will need to be dressed on an as-needed basis—in any event, at least twice a day. So I will be trained this afternoon for my new role as *Nurse Susan*. I must say that the patient has a decided lack of confidence in my capabilities, although the doctor and the (real) nurse say that it will be easy and that I really cannot hurt him.

SARATOGA SPRINGS

*No man ever committed suicide with
an unraced two-year-old in the barn.*

—Racetrack aphorism

BETWEEN ENGAGEMENTS

POSTED BY CHARLIE:

This morning, I had appointments with Dr. Jarnagin and Dr. Saltz. Dr. Jarnagin removed the remaining thirty-six of the original forty staples—four had previously been removed to treat the post-surgical infection that developed. Dr. Jarnagin also scheduled for August 3 the surgery to remove the right lobe of my liver. Dr. Saltz indicated that he will probably restart my chemotherapy a few weeks after the August 3rd surgery.

Dr. Jarnagin indicated that the trauma and pain associated with the upcoming surgery should be comparable to what I just experienced. Just a day or so ago, the pain from the surgery on June 15 began to subside. I am starting to sleep a bit better and hope soon to start getting through the nights without the assistance of a nurse's aide. Yesterday, I was able to walk for the first time with a fairly normal stride and posture. Today, I began to cut back on the pain pills. A visiting nurse service is coming twice a day to change the bandages on the infected wound, and I am scheduled to continue taking antibiotics for approximately another week. Although my belly is still distended—I am wearing size double extra large sweat pants—the swelling is starting to subside. As soon as I have reduced my ingestion of pain medication enough so that I am not groggy, I will post an update on this blog inviting visitors.

ON TRACK

Gradually, my energy and stamina seem to be improving. I am no longer taking medication for pain or antibiotics for the infection of the incision. Although the infected part of the incision is slowly healing, it still requires the services of a visiting nurse twice a day. The swelling of my abdomen has subsided enough to enable me to wear newly purchased khaki trousers only four inches larger in the waistline than I wore prior to the surgery. The skin below the incision was irritated from something during the surgery, and that skin irritation does not seem to have lessened. I will have to find out if Dr. Jarnagin can safely modify the next surgical procedure to avoid a recurrence of that skin problem. Next Tuesday, I will undergo a day of tests to ascertain that I am on track for the second surgery to take place as scheduled.

Starting tomorrow, I am going to start seeing visitors again. I will not be able to do much during such visits. I have been taking walks of a half mile or so, with my longest walk not exceeding 1.4 miles. Most foods no longer taste good to me, apparently because of what is transpiring with my liver as its right lobe atrophies. But I am looking forward to being with friends again, if only to talk.

HALFWAY POINT

It has been twenty-four days since the surgery, and the next surgery is scheduled twenty-five days from now. Candidly, I am dismayed by how weak I find myself. Yesterday, I had lunch at my apartment with my brother, Bill, who drove up from Virginia to see me, and I was so exhausted afterwards that I had to take a nap. Today, I went out to a local restaurant for lunch with a friend and got so tired that I developed an upper back spasm, which thwarted my plan to be driven out this afternoon to Belmont Park to see my two-year-old colt, Backslider. Backslider arrived from Kentucky two days ago to be prepared for his racing debut on July 23 at Belmont Park.

It is amazing how fast, and to what extent, a sequence of cancer, chemotherapy, surgery, and infection can debilitate one's body. Since the surgery, I have lost nine pounds, and I would not be surprised if all of the weight lost has been muscle mass. At least the post-surgical swelling of my abdomen has receded enough so that, when I went out to lunch today, I was able to wear a pair of trousers that fit me before the surgery.

I need to get much stronger over the next twenty-five days before undergoing the upcoming surgery. In the meantime, I will try to husband my strength enough to enable me to go out to Belmont in the next couple of days to see Backslider. After I had to cancel my visit today, Christophe kindly sent me two photographs of Backslider. Obviously, Backslider is still growing, as his hindquarters are higher than his withers. But he looks strong, and my guess is that he will mature into a grand-looking colt.

JULY 12, 2009
SUNDAY 6:30 PM

UP AND ABOUT

After a wet, cold spring, the last few days in New York City have been sunny, with low humidity, and temperatures not exceeding around eighty degrees. I had enough energy on Friday to visit Belmont Park to see the two horses that we have stabled there. On Saturday, I went out to lunch with a friend who drove into

Manhattan to see me. Today, I took three walks totaling, according to my pedometer, 2.45 miles.

Since Friday, the nursing service has had to come only once a day, instead of twice a day, to dress my infected incision. The incision has taken much longer to heal than the surgeon's optimistic estimate. With luck, it will heal not only before my next surgery, but in time for me to go up to Saratoga Springs for a few days prior to the surgery.

JULY 14, 2009
TUESDAY 6:55 PM

SEARCHING FOR FRANKL

The feature article by Melinda Beck on page D1 of today's "Health Journal" section of *The Wall Street Journal* is entitled, "A New View, After Diagnosis: Experimental Group Therapy Aims to Help Cancer Patients Find Meaning in Face of Mortality." The article discusses an experimental program at MSKCC aimed at helping "cancer patients find a sense of meaning, peace and purpose in their lives, even as the end approaches." The program is based in part on the writings of Viktor Frankl, specifically *Man's Search for Meaning*. I count myself as fortunate that I had read this book and therefore was aware of it as a resource, prior to my diagnosis.

Today, I spent the day at MSKCC having my liver and general condition tested and examined. It seems that the right lobe of my liver is atrophying, and the left lobe of my liver growing, as planned. Moreover, even though I am appalled at my loss of condition this calendar year, the doctors and nurses seem to think that I am rebounding exceptionally well from my surgery. The tentative August 3 date for my next surgery is now a firm target.

The doctors and nurses cannot figure out what caused the skin around, and particularly below, my surgical incision to become hypersensitive, then peel and become pink and itchy. One guess is that the trauma of the surgery triggered a mild case of shingles—a mild case because I had been vaccinated for shingles. Whatever caused this skin condition, it seems to be running its course and is now only a minor irritation. I am hoping that the next surgery will not cause it to flare up again, as a mere touch of cloth to my abdominal region was difficult to bear, which interfered with my sleep and made standing up while wearing a shirt a dreadful experience.

JULY 16, 2009
THURSDAY 7:09 PM

MY MOTHER'S BIRTHDAY

Today is my mother's ninety-fifth birthday. She and my father and sister are going out to dinner to celebrate at the Ponte Vedra Club.

Susan and I and a friend went out to Belmont Park today to watch Hot Money run in a stakes race restricted to three-year-old horses bred in New York. He was the 4-to-5 favorite in a field of nine. He finished next to last.

Yesterday, I received a call from MSKCC that the radiologist reviewing the CAT scan that I underwent the previous day had detected a pulmonary embolus—a blood clot—caught in the IV filter in my vena cava. This blood clot is almost certainly caused by my cancer. Such clots must be prevented, as they cause heart attacks and strokes if they get by the IV filter. Starting yesterday, I am now injecting myself once a day with a blood thinner, a low-molecular-weight heparin called dalteparin (Fragmin). For a while I will have to have my blood platelets checked twice a week at MSKCC. Starting today, I have to wear compression hosiery, which I find to be hot and uncomfortable. (I can't get them to stay up. Maybe they're the wrong size, or maybe I need a garter belt?) Because I am now on a blood thinner, I can no longer take anti-inflammatory medications like NSAIDs (aspirin, ibuprofen, etc.) or Celebrex to control the arthritis in my left hip. A primary concern for me is whether either the blood clots or the blood thinner will preclude certain chemotherapy options. In any event, I will now have to be treated for prevention of blood clots for a minimum of six months. If I were no longer deemed to have active cancer, such treatment could be discontinued.

JULY 18, 2009
SATURDAY 6:12 PM

IT'S AN ILL WIND THAT BLOWS NO GOOD

For the first time this year, I visited the glaucoma specialist whom I had been seeing every two months for eight or nine years. Interestingly, my intraocular pressure was the lowest that it has ever been (that's a good thing), and my visual field test seemed to indicate somewhat less damage than previously (which I'm not sure is possible, as dead neural ganglion cells don't revive or get replaced). Glaucoma is thought by at least some specialists to be an autoimmune disease, and chemotherapy suppresses the immune system. Thus, chemotherapy may be efficacious for glaucoma—or, at least, my chemotherapy regimen may be beneficial for my particular type of glaucoma. Until I got cancer, my only material ailment was glaucoma, which is to some extent heritable and from which my mother is now legally blind. In addition, my father suffers from wet macular degeneration. Prior to my diagnosis of incurable cancer, I had worried about outliving my eyesight.

Over the last few days, although I am still instructed not to lift anything over five pounds in weight, my strength and stamina seem to be returning. Today, I took advantage of beautiful weather and walked over five miles, including two separate walks of two and a half miles along the pathway paralleling the East River. It is a nice walk, as I have to cross only two through streets. A highlight of the walk is passing by an enclosed dog run, in which the dogs always seem to be frolicking together. I suppose they must fight with each other sometimes; but as I walk by, I haven't seen or heard any signs of quarreling.

GOLF'S GOOD GUY

Tom Watson is the only top tournament golfer that I have met. Eleven years ago, the night after he won his last PGA tournament, the Colonial, I was seated next to him at a dinner in New York at the Links Club. He was the guest of honor. It was considered a remarkable feat for him to have won a PGA tournament at the age of forty-eight. I found him to be a good dinner companion—dignified, personable, and interested in more than just himself and golf. I was not surprised that he conducted himself so well today after his crushing failure to maintain his lead—at age fifty-nine, with a recent hip replacement—on the final hole of the British Open. I could not help comparing and contrasting his demeanor with Tiger Woods's swearing and pounding the tee box with his club after hitting a bad shot.

BAD HABITS DIE HARD

In getting exercise, when I act both as the athlete manqué and as my own trainer or coach, stupidity tends to prevail. My instinct is to push constantly for personal bests, even though I try to tell myself to go slowly and not experience a setback by injuring myself. On Saturday and on Sunday morning, as part of each of the three two-and-a-half–mile walks that I took, I climbed up and down a staircase with fifty-seven steps. I put aside the knowledge that I am still very weak from my June 15th surgery, and I mused about whether I should try to get each walk up to five miles before my surgery on August 3.

Now I am nearly immobilized by pain in my lower back, which I am not permitted by the doctors to treat with anti-inflammatory medications. I hope that this self-inflicted problem does not compromise my stamina too much heading into the surgery or leave me to deal with a combination of abdominal and back pain after the surgery. I hope also that the back pain does not curtail either my ability to see people before the surgery or our plan to spend a few days in Saratoga Springs.

SARATOGA SPRINGS

This morning, after having my blood-platelet level checked at MSK to see how I am reacting to the daily injections of blood thinner, Susan and I drove up to Saratoga Springs. The racing starts Wednesday and continues through the Labor Day weekend. I have been coming to Saratoga for the racing since 1963, and Susan started coming here with me a few years later. There are more important races at

Saratoga than there are at any other race meet of comparable duration. Saratoga also features races for two-year-olds. I enjoy going to the walking ring before the maiden races (i.e., races for horses which have never won a race) for two-year-olds. I try to identify future stars among the horses that have not yet raced, and I study the characteristics of the progeny of young stallions.

Just before my cancer diagnosis, we had rented an apartment in Saratoga for the entire racing season. One of the things that we had wanted to do when I retired was to spend the racing season here. In addition to the horse-related activities, there is plenty to occupy your time in Saratoga. When I was working, we drove back and forth between New York City and Saratoga on weekends. I also usually took a vacation here for the entire week during which the yearling auctions were conducted. Typically, I would bid on several yearlings and buy one or two.

This year, we can stay in Saratoga only through August 1, as I have to get back to New York City for my surgery on August 3. Although I am still limited in my activities by back spasms and dopey from the medicine that I am taking for the back spasms, we will make the best of these next few days.

FOUR DAYS OF RACING

Although I lost a modest amount of money on the races, and there was considerable rainfall until today, attending the first four days of racing at Saratoga was a tonic for me—appropriately enough, as Saratoga Springs was a spa long before the racing began in 1864. There were some good races all four days, with the outstanding card today featuring the Grade I Diana on the turf, a race for fillies and mares; and the Grade II Jim Dandy for three-year-olds being prepped for the Grade I Travers, the "midsummer Derby," four weeks from now.

In the midst of the chronic economic depression of upstate New York, the town of Saratoga Springs has become more prosperous and gentrified. Horse owners have restored Victorian mansions; developers have constructed, and are still constructing, attractive condominiums (I do not know if they are finding buyers); chain stores have mixed in with the mom-and-pop shops on Broadway; and, although the proprietors of Saratoga's best restaurant, Chez Sophie, are closing it and moving to France, there are more good restaurants in and around town than ever.

In spite of the gentrification, Saratoga Springs retains its magic. The town not only has the natural advantage of its springs, but also a setting amidst lakes and rolling farmland. The perennial success of the race meet defies all logic, especially in the context of the accelerating decline of the sport of thoroughbred racing. Saratoga is located far from major metropolitan areas; the clubhouse and stabling facilities are antiquated; there is a severe shortage of hotel and motel rooms, parking, and seating; the horsemen have to pay inflated rents for housing; and on the days when it isn't raining, it is often hot and humid. Nevertheless, like most horse owners and fans of racing, there is no place I would rather be from late July through Labor Day.

With the exception of some California owners and trainers who go to Del Mar, California, most of the leading figures in American racing come to Saratoga for at least part of the six-week meet. Days on the racetrack start before dawn, and Saratoga offers nonstop social life, at all levels, during those six weeks. Daily life offers summertime small-town charms, such as the farmers' market. For thoroughbred horse-industry professionals and insiders, the focus of activity, including conferences and yearling sales, continually shifts throughout the meet. During a given week, one might find the steeplechase crowd in town, or the Kentucky breeders and yearling consigners, or the coach-horse people. There are families of owners, breeders, and trainers that have been stalwarts

of racing and have come to Saratoga for decades. There are new owners who have made a lot of money in the latest hot area of the economy, most of whom disappear from racing and Saratoga in two or three years—a few of whom endure to become part of the fabric of the industry.

Although I would like to meet Susan's goal of getting back to Saratoga before Labor Day, I cannot set such an objective. I simply have to follow the cliché of taking each day as it comes. Following my surgery on June 15, it was predictable that I would be weakened and lose weight, but it was not predictable that I would develop shingles, an infection of the incision, a blood clot, and muscle spasms of the lower back. In addition to the painkillers, the various drugs that were administered to me for these complications also robbed me of energy and clarity of thinking—as reflected in the paucity of my blogging since the surgery. (After I got off the last of these drugs two days ago—other than the blood thinner with which I will have to continue injecting myself for at least six months, I was relieved to find that my handicapping of the races improved markedly yesterday and today.)

To deny that I am facing this coming surgery with trepidation would be transparent bravado. With my memory of the last surgery so fresh, I have no illusions about the suffering that is in store for me, even if all goes well. Debilitated not only by lingering effects of the four complications that developed after that surgery, but also by loss of weight, muscle mass, and stamina, I go into this surgery with much less strength in reserve. The surgeon seems to think that I am doing relatively well and that I am capable of staying on an ambitious schedule. Apparently, some patients have to wait months between the two surgeries. I keep reminding myself that for the first two months after my diagnosis, I was told adamantly that surgery was not an option for me and that my disease was incurable.

Having undergone these two surgeries, I will have much better odds of surviving for a few more Saratoga race meets, and even some chance of being cured. I approach Monday morning's surgery keenly aware of how lucky I am to be going into it with strong support from family and friends. Susan plans to communicate with you by posting on this blog after the surgery.

BACK IN ROOM 1937

A hospital is no place to be sick.

—Samuel Goldwyn

SURGERY UNDER WAY

POSTED BY SUSAN:

While waiting in the lobby, David and I just received our first report from "Roseanne, the Patient Liaison Representative," whose job is to report every two hours to the families waiting for patients in surgery. We learned only that the procedure got under way at 9:50 a.m. Dr. Jarnagin previously told me that the surgery will take about three to four hours.

ROSEANNE'S REPORT

POSTED BY SUSAN:

Roseanne just made her rounds of the lobby and reported that Charlie is tolerating the surgery well and that the surgeons are now at the stage of "clamping off the arteries and blood vessels." She said that the procedure will probably take another hour, but that she doubted that she would see us on her 3:00 p.m. rounds.

ONCE AGAIN, NO SURPRISES

POSTED BY SUSAN:

Charlie is in Recovery, and Dr. Jarnagin reported, as he did after the last surgery, that there were no surprises—no cancer in other parts of his abdomen, the left lobe of the liver looked as expected following the resection of seven weeks ago, and the removal of the right lobe went as planned. Charlie will remain in Recovery overnight—until perhaps 5:00 a.m. tomorrow morning—as is standard with this type of operation. David and I will see him for a five- or ten-minute visit in another hour. Meanwhile, we are going to get some lunch.

FAVORABLE SIGNS

POSTED BY SUSAN:

David and I had a short visit in the Recovery Room with a groggy Charlie, but there were signs that he was still himself. David told him that there are reports from the New York Giants training camp that Eli Manning is throwing flutter balls, and he gave an exasperated sigh—one of his favorite contrarian theories being that the Giants would be much better off with a different quarterback. He immediately asked for his glaucoma drops and supervised the nurse's administration of them. And, of course, he smiled upon hearing Dr. Jarnagin's report on the surgery.

We are home now and will go back for another ten-minute visit in the Recovery Room about 8:00 p.m. The nurse told us that it takes longer after liver resection surgery than after other surgeries of comparable length for the anesthesia to wear off because of the role of the liver in metabolic functions.

FIRST MORNING

POSTED BY SUSAN:

Charlie was moved from the Recovery Room to his room this morning at 9:00 a.m. Because the level of nursing care in the Recovery Room is more intense, he was kept there for the full night—until all lab reports were good. He is comfortable (he just rated his pain for a nurse at "1" on a scale of 1 to 10), and David is reading him news reports on horse racing and New York Giants football. He will be getting up to walk within the next hour. He is concerned that the back spasms he experienced in the last two weeks will interfere with his setting new distance records doing laps in the hallway, but the staff is certain that it will be able to control the spasms should they be a problem. All in all, he is doing well.

AFTERNOON REPORT

POSTED BY SUSAN:

A few months ago, I was warned that there are a lot of "curves in the road" with cancer and it is best not to take any of them too fast. I was advised not to get too high when there's good news, and not to get too down when there is bad news. Today has been such a good day that I am having to try to remember not to get too elated.

Charlie looks, feels, and acts better than he did at this stage after the first surgery. We are guessing that the shingles that he might have developed after the last surgery took more of a toll than we realized. It seems easier to manage the pain this time, and he is able to walk without bending over to avoid cloth touching the skin under the incision.

When Vivian, the physical therapist with whom Charlie had great rapport during the last hospital stay, saw that he was in the hospital again and due for therapy, she immediately signed up for his case. She has presented him with a new challenge, although I could see that she was alarmed at her patient's level of enthusiasm. This afternoon she timed him with a stopwatch to see how many hall laps he could do in six minutes. Two and one-quarter laps is now his six-minute personal best. So, in addition to working toward maximum mileage, he now has a speed goal.

One member of Dr. Jarnagin's team was here this afternoon, and we are expecting the whole team this evening. I hope that they will confirm medically what I am observing—that Charlie is recovering remarkably well.

AUGUST 5, 2009
WEDNESDAY 9:59 AM

LOOKING WELL

POSTED BY SUSAN:

Charlie continues to look and feel well this morning. He is experiencing much less pain than he did after the last surgery. As a result, he is using his morphine drip button less often, which makes his mind clearer and his ability to enjoy the newspapers and his e-mail greater.

Not surprisingly, Charlie's liver function is not yet normal, but the doctors are not concerned. They say that situation is expected and should resolve in a few days. Dr. Jarnagin's nurse practitioner told Charlie this morning that, in her experience, patients often have a decline in energy on the third day, and he should anticipate that possibility and not be discouraged.

Charlie is walking the hallway with the respiratory nurse. (He walked ten laps, 0.71 miles, yesterday.) As with all surgeries, there is a lot of focus on keeping his lungs clear, so his diligence with the walking and spirometer is important.

SPIRITS LIFTED

POSTED BY SUSAN:

Charlie was started on antibiotics this morning to combat cellulitis around the incision, and he became quite discouraged. As he increased his usage of the patient-controlled pain medication, he became groggy, and to David and me, he did not seem as healthy as yesterday.

But Dr. Jarnagin and his full team were just here, and Charlie's spirits have improved tremendously with the news that his liver function is returning to normal, the suggestion that he needs to remember how far he has come, and the characterization of the cellulitis as hardly a "self-respecting infection." So, he has gone back to the hallway to walk more laps.

WITH DAVID

POSTED BY SUSAN:

Charlie is not always comfortable, but his medical progress is continuing. Dr. Jarnagin is pleased with how he is recovering. The increased pain and discomfort relate in part to the normal, expected buildup of fluid and gas in the abdominal area. (I found it interesting that he has gained about twenty pounds since entering the hospital, while eating and drinking nothing but clear liquids until today. Obviously, he is retaining a lot of fluid.)

David and I continue to trade off shifts from time to time at the hospital, but he has done more than his share today, allowing me to get over to the gym and have lunch at home with our cat, Pepper. In addition to David's help at the hospital, I derive great comfort from having him with me during this stressful time.

AUGUST 8, 2009
SATURDAY 12:15 PM

COMING ALONG

POSTED BY SUSAN:

Charlie says that he is feeling better than yesterday and a lot better than the day before yesterday. He is allowed to eat anything, but he has very little appetite and has been content with a little applesauce or a banana. The Fellow came early this morning, so I did not speak to him myself, but I understand that he reported that everything is coming along fine. The nurse also said that Charlie is doing well.

Charlie had a (painful, because of the positioning) CAT scan last evening in connection with Dr. Jarnagin's research study, and his liver looks as expected. He continues to have an intravenous tube connected to his Mediport, which would seem to indicate that he will be in the hospital at least a couple of more days. He is sleeping a lot, and his activities have been pretty much limited to the highly recommended walking and sleeping. He did not do much reading after the last surgery either, but the amount of sleeping is much greater this time around. I believe that he is not as uncomfortable as he was the last time, which is allowing him to get more rest.

We are all looking forward to television coverage this afternoon of both golf and horse racing. We are rooting for our trainer Christophe to win the Arlington Million with Gio Ponti and the Secretariat Stakes with Laureate Conductor. These important races will be televised on ESPN.

JUST WHERE WE WANT HIM TO BE

POSTED BY SUSAN:

Charlie's IV came out today, and he is being switched to oral medications. Tomorrow morning, the doctors will evaluate whether he can go home later in the day, but they have indicated that they are in no hurry and will see how he feels. He seems stronger today, has a better appetite, and is able to concentrate a bit more. The Fellow said he is just where they want him to be at this point.

It was fun to be able to watch horse racing yesterday on ESPN. Although New York City cable television has a horse-racing channel, the hospital does not carry it. In fact, we cannot even see horse racing on the computer because we use the hospital-provided Wi-Fi, and any website with a connection to gambling (or to pornography or social networking) has been blocked, apparently to keep the employees focused on their jobs. Yesterday's racing was especially exciting for us because we were thrilled to see Christophe's horse win the Arlington Million. Years ago, it was established as the first race to offer a total purse of $1,000,000. The winner gets $600,000 of the purse, and the trainer's fee, here to Christophe, is twelve percent of the winning share, or $72,000.

AUGUST 10, 2009

MONDAY 3:13 PM

HOME AGAIN

POSTED BY SUSAN:

Charlie was discharged early this afternoon. He is feeling much better today, and the doctors are happy with his medical progress. He is definitely more comfortable than he was at this stage after the last surgery. I am again being allowed to hone my nursing skills, as the incision is still draining and needs redressing frequently.

Being out of the hospital seems to have lifted Charlie's spirits. He is even allowing himself to think that we may make it back to Saratoga toward the end of the meet.

AUGUST 12, 2009

WEDNESDAY 10:10 AM

196 POUNDS

POSTED BY CHARLIE:

Following my surgery on June 15, I recovered sufficiently to meet the surgeon's criteria for withstanding removal of sixty-eight percent (a bit more than he may have anticipated) of my liver on August 3. On June 14, I weighed approximately 172 pounds, five pounds or so more than I might weigh ideally to get comfortably into most of my trousers. On August 2, I weighed 161 pounds. Since the surgery on August 3, my weight has ballooned as high as about 196 pounds and was 193 pounds this morning.

I have two primary near-term objectives: to stay home and to start thinking more clearly. The medical team is working not to readmit me to MSKCC to deal with cellulitis and with edema. Meanwhile, I am struggling to regain enough power of concentration to hold a thought firmly in my grasp from the beginning of a sentence to its end.

STILL ON THE JOB

POSTED BY SUSAN:

I thought my writing days might have ended, but Charlie finds using the computer very tiring, so I will be posting for a few more days. The recovery is definitely following an upward trend, but it is not a straight line. Some days are better than others, and there are always issues with which he must deal with the help of the medical staff. Such issues seem to be a normal and expected part of recovering from major surgery, and the nurse practitioner answering calls this week is kind and helpful.

Today is a good day. Charlie has been out four times to walk around the block, about a quarter of a mile each time. Overall, he is feeling much stronger and more comfortable than yesterday.

ROOM WITH A VIEW

POSTED BY SUSAN:

Last night, Charlie began to have a fever, and when I changed the dressing on his incision, there were signs of infection. We ended up back at Urgent Care. He was seen almost immediately by a surgical fellow, and, unfortunately, when staples were removed, the incision proved to be infected. More staples were removed this morning, so the incision is open and packed along almost its entire length, and Charlie is again on IV antibiotics. His temperature is back to normal.

After seeing the doctor, we spent another four hours in Urgent Care while waiting for a hospital room and then for Patient Escort to take him to it. During the last hour (1:00 a.m. to 2:00 a.m.), Charlie, David, and I strangely became a little silly, and the two of them began accusing me of optimistic bias in my blog postings and suggesting how I might spin this latest development:

"His temperature dropped to only ninety-nine degrees within two hours of the incision being cleaned."

"He was pleased to see one of his favorite nurses, Marilou, when he arrived on the nineteenth floor."

"Only the exclusive nineteenth floor had private rooms available, and Charlie has a beautiful view of the Manhattan skyline and room decor by the Four Seasons hotel chain."

All of the above is true.

I do not want to make light of a bad situation. It is disappointing to be back in the hospital. On the other hand, I have a feeling that everything is under control. The nurse is trying to have the Wound Care specialist team see Charlie later today. He looks well and was up walking the hallway before 7:00 a.m. this morning.

FRIDAY 2:38 PM

THE WHEELS ARE NOT OFF

POSTED BY SUSAN:

As you have probably noticed, Dr. Jarnagin and his team are prone to clichés, and the latest is that Charlie's current situation is a setback, but the "wheels have not come off." We are also told that they can "fix this."

Charlie has a lot of fluid in his abdomen (seen on a CAT scan), and the fluid is most likely infected. This afternoon an interventional radiologist will insert a drain, remove as much fluid as possible, and leave the drain in place, where it will remain for more than a week. He should start to feel a lot better pretty quickly after that. The incision will be cleaned and repacked a couple of times a day for a while.

The liver looks "great" on the CAT scan—a lot larger—and most of his blood work also looks good. The liver is functioning well. His protein is low, but as the infection diminishes, better appetite and nutrition should correct that.

Of course, we are concerned and disappointed, but we are told that this surgical team deals with this situation often when it has performed aggressive treatment procedures. Charlie is getting a lot of attention and has had multiple visits from members of the team. I have confidence in them. I will post again tomorrow, unless there is significant new information later today.

AUGUST 15, 2009
SATURDAY 9:40 AM

MAD MEN

POSTED BY SUSAN:

The drainage procedure took place yesterday afternoon under sedation and local anesthesia, so last evening Charlie was groggy and on bed rest. The nurse says that there is a lot of drainage, which is what the doctors want.

This morning, Charlie is feeling better. He ate a large breakfast, and, when walking in the hallway, was more erect and speedier than he has been. His temperature is normal. Someone from the wound care specialist team will be visiting him this morning.

The doctor came on his rounds very early this morning, so I missed him and was not able to speak with him. After his evening rounds, I will write more.

Today's planned activities, for Charlie, David, and me, include watching the last episode of season two of the *Mad Men* television show in preparation for the start of season three tomorrow. We have a DVD player, a flat-screen TV, and a Bose CD player in the room.

UNDER CONTROL

POSTED BY SUSAN:

The Fellow just made his evening rounds, a bit before evening because his family is arriving from Ottawa, Canada, and confirmed what seems obvious from observing Charlie. He is much better. The drain is removing copious quantities of liquid from the abdomen, and the blood work is all fine. The culture of the drainage material was negative, but that was predicted, because Charlie was already on antibiotics.

Charlie's appetite is excellent, he looks well, and I am feeling positive. The wound care team plans to send him home next week with an automated vacuum dressing that needs to be changed only every three days. Charlie and I were both impressed with the size of the wound when it was fully opened up today for cleaning, but the team said that it looks good.

AUGUST 16, 2009
SUNDAY 1:16 PM

SURGEON'S VACATION

POSTED BY SUSAN:

There is not much to report today. The Fellow covering for Charlie's Fellow made a brief visit before I arrived. (Dr. Jarnagin has left on vacation, but is available to the Fellows by phone.) Apparently the covering Fellow inquired as to how Charlie feels, which is quite well, but offered no opinions or information. The nurse says that the wound looks good. The drain is still producing fluid. The diuretic is beginning to increase Charlie's urination, in an effort to reduce the estimated thirty pounds of excess fluid in his body. Charlie's appetite is good, and he is eating very well.

PREPARATIONS FOR DISCHARGE

POSTED BY SUSAN:

The current plan is to discharge Charlie on Wednesday. One high-tech and one low-tech treatment will be ongoing at home, and the staff here at MSKCC is helping us make plans to manage those treatments. The high-tech treatment is called VAC (vacuum-assisted closure), or vacuum therapy. It uses a battery-operated device that applies sub-atmospheric pressure to the incision wound through a specialized wound dressing. This therapy promotes wound healing, and the dressing needs to be changed only three times a week, rather than twice a day. Charlie is lucky that his wound is amenable to this therapy. The speed of healing when using it can be dramatic.

The low-tech (or maybe just *lower-tech*) treatment is the drain that is continuing to remove fluid, albeit at a much reduced rate than previously, from the abdomen. The drain will be removed at the earliest next Tuesday, and it could be needed for some time beyond then.

Charlie's temperature remains normal, and the scheduled discharge is obviously an indication that he is making progress. He looks well but is very fatigued from this ordeal. So am I.

APREHENSION

POSTED BY SUSAN:

Although I am sure that Charlie will enjoy being home—sleeping in his own bed, eating home-cooked food, and reading in a comfortable living room—he is apprehensive about the discharge tomorrow. The visiting nurse will come only three times a week, and that is the only planned contact with a health professional until next Tuesday. His confidence in my nursing skills has not risen, and I will be in charge of the drain twice a day. (My duties actually relate more to those of a plumber than of a nurse.) Despite the problems with hospital life, it is comforting to know that a trained person is only a few steps away.

An aching back is an additional contributor to a less than cheery Charlie. The Fellow has ordered an ultrasound, just to make sure that the problem is musculoskeletal. The nurse practitioner says that everyone complains about the uncomfortable beds here at the hospital, so we all suppose the back problems relate to the bed and to moving about without full use of the abdominal muscles.

With luck, I will be writing a more upbeat post from home tomorrow. As I see it, we have some inconveniences, some things with which we must deal, but the health picture is much better than it was four days ago, and monumentally better than four months ago.

THE VISITING NURSE

POSTED BY SUSAN:

We finally left the hospital in the early afternoon. I had arrived at 7:00 a.m. for the scheduled early departure, but a couple of snags kept us there in ill humor until 1:00 p.m.

Amazingly, things brightened considerably with the arrival of a nurse from the Visiting Nurse Service of New York (VNSNY). A few of our previous contacts with VNSNY had not made us especially hopeful for anything beyond a person with technical competence to use the VAC wound therapy machine. The nurse assigned to Charlie's case, Phil Leon, proved to be knowledgeable, kind, helpful, and professional. He demonstrated his "can do" attitude when we found that the box of supplies provided by the VAC machine rental company did not include a crucial type of wound dressing. Phil called another client and arranged to go to his home and borrow what was needed to install the VAC today and then returned to finish the job. He gave some great suggestions for getting more protein and calories into Charlie's diet, and he volunteered to arrange for in-home physical therapy to help with the back pain. Charlie and I both felt relieved and optimistic when he left. Phil is scheduled for visits three times a week, but he offered to stop by tomorrow to check on how Charlie is doing and to see if there are any problems with the VAC.

A CHALLENGE FOR CHARLIE

POSTED BY SUSAN:

When I am working with my trainer at the gym and tell him that something I am doing is hard, he always replies that the exercise is a challenge for me. Evidently, a trainer learns never to allow a characterization of *hard* on the theory that *challenge* is more positive and will keep the client motivated. Keeping this in mind, I would say that being at home with the drain and the wound VAC is a challenge for Charlie, and for me as the primary "caregiver." Nonetheless, Charlie seems to be making gradual progress. The interventional radiologist will evaluate where we are with the drain on Monday. Charlie definitely looks better to me, and he is now out for his third four-block walk of the day.

This afternoon, we watched some of the Saratoga horse races on television. Charlie and David are looking forward to the Giants' preseason game against the Bears this evening.

HOORAY

POSTED BY SUSAN:

Charlie and I spent the day at MSKCC, and all the news was good. An ultrasound and blood work showed that his kidney function is fine, so the diuretic dosage could be quadrupled to help rid his body of the thirty or so pounds of excess fluid he is carrying in his lower body. The remaining staples were removed from the incision; the removal was not as painful as it was after his previous surgery, and the infection is healing well. (MSKCC's staff estimate that the wound VAC has already saved him two to three weeks of healing time versus the old-fashioned dressing.) His abdominal cavity has dried out, and they removed the drain—no more plastic bag to pin inside his clothing when he goes out. An ultrasound examination of his legs revealed no new blood clots. All of his blood work was favorable.

We feel pretty good tonight. It looks as if Charlie has turned the corner after the infection detour. (Have I picked up Dr. Jarnagin's habit of using clichés?)

AUGUST 28, 2009
FRIDAY 12:28 PM

BACK IN ROOM 1937

POSTED BY SUSAN:

We had a regular appointment with Dr. Jarnagin this morning, and he recommended hospitalizing Charlie again for a couple of days to treat his continuing edema and other related issues more aggressively than is possible at home. He will also receive a transfusion for the anemia that he has had ever since last March. Unlike previous returns to the hospital characterized as "setbacks," "detours," or "bumps in the road," this one is supposed to put him on a faster track to full recovery and to make him more comfortable. Continuing to carry thirty-plus pounds of excess fluid is unpleasant, so Charlie is actually happy to be readmitted. By coincidence, he is in the same room that he occupied last time—Room 1937.

MINUS SEVEN AND COUNTING

POSTED BY SUSAN:

The more aggressive treatment is producing results. Charlie has lost seven pounds since he entered the hospital yesterday. He is not yet comfortable, but the loss is encouraging. Being in the hospital where his blood profile can be tested as often as needed, the doctors are able to give him larger doses of the diuretic Lasix. (Lasix is widely used in horse racing and is often thought to be performance-enhancing, so this regimen has been the subject of a few bad family jokes.)

Another good medical development is that the incision wound has healed to such an extent that use of the wound VAC is no longer necessary or even feasible. The wound is well on the way to being completely healed. An additional bonus is that Charlie no longer is attached to any external equipment.

DRAINING FLUID

POSTED BY SUSAN:

The treatment continues to be the same, and Charlie lost a couple more pounds on the scale. As seen in the frequent blood tests, he is tolerating the Lasix well. If he continues to tolerate it, the doctors may increase the dosage again tomorrow. He is still uncomfortable. While his legs no longer look elephantine, it would be hard to say that they are anything but huge. In answer to my question this morning, the Fellow said that Charlie would probably remain in the hospital until most of the fluid is gone.

DOWN THIRTEEN

POSTED BY SUSAN:

There is nothing much in the way of news today, and that is a good thing. As of this morning, Charlie had lost thirteen pounds since entering the hospital on Friday. The doctors switched the Lasix delivery today to oral from intravenous in anticipation of returning home within the next couple of days. The wound looks better and smaller. On the negative side, he continues to have a considerable amount of discomfort and cannot find a completely pain-free position.

HOMESTRETCH

POSTED BY SUSAN:

Tomorrow will be the one-month anniversary of the second liver-resection surgery. Although it has been a difficult month, and we did not expect that Charlie would mark that anniversary as an inpatient at MSKCC, I feel that we are now in the homestretch.

The most important things are that all known cancer has been removed, and the liver is regenerating. The infections have been cleared up, and the excess water is coming off. He weighs about nineteen pounds less than when he entered the hospital last Friday.

It appears that Charlie does not process the oral version of Lasix—there was virtually no progress when he was switched over from the intravenous version, and he has now been switched back. That is the main reason he is still in the hospital.

Having painted the rosy picture above, it is important for me to report also that he faces probably four months of chemotherapy in an attempt to kill lingering, undetectable cancer cells, and that will be difficult. Also, he is still not completely comfortable, which has disturbed his sleep for weeks. He is exhausted.

SEPTEMBER 3, 2009
THURSDAY 4:29 PM

GOOD FOOD, LOTS OF SLEEP

POSTED BY SUSAN:

Charlie came home today weighing twenty-two pounds less than when he entered the hospital a week ago. He is out right now enjoying a walk in today's lovely weather, and we walked home from the hospital in the early afternoon when he was discharged. What I think he needs most right now is good food and lots of sleep. I will make every effort to see that he gets both.

SEPTEMBER 10, 2009
THURSDAY 5:15 PM

OUT TO DINNER

POSTED BY SUSAN:

Not everyone has been sure that "no news is good news," so I thought that I would write a brief post to say that there has been some progress—a little more strength, a little less pain, and a lot less bloat. Charlie's weight is back to what it was before the surgery, although he has clearly lost muscle mass and is still carrying excess fluid, primarily in his feet. Tonight we will see if the Italian restaurant Raffaele, on First Avenue near 57th Street, can fatten him up a bit. It will be the first time that we have been out since the beginning of August.

NUTRITION BE DAMNED

POSTED BY SUSAN:

Yesterday, Charlie had a follow-up visit with the surgeon, Dr. Jarnagin (I was playing golf), and later in the day, I joined him for a meeting with Dr. Saltz. Both doctors believe that Charlie needs more time to regain strength before restarting chemotherapy. He will be re-evaluated in two weeks with a probable start date for the chemo at the beginning of October.

In the meantime, his instructions from Dr. Saltz are to eat a calorie-laden diet of high-fat foods (nutrition be damned) and to do his best to get an hour of aerobic exercise a day. Given his poor appetite, it is not clear which assignment will be more difficult. I have visions of us appearing for the appointment in two weeks with me the one who has gained ten pounds.

Charlie is facing the upcoming chemotherapy with apprehension. Dr. Saltz said the regimen will be changed to the one known as FOLFOX from the FOLFIRI he had before the surgeries. FOLFOX's set of frequent side effects is worrisome. In particular, FOLFOX often causes numbness of the fingers, and Charlie already suffers from Raynaud's syndrome, which means that he risks losing the use of his fingers, perhaps permanently, from taking FOLFOX.

WHY ME?

*In my happier days I used to remark on the aptitude
of the saying, "When in life we are in the midst of death."
I have since learnt that it's more apt to say,
"When in death we are in the midst of life."*

—A Belsen Survivor

OUT OF A MIASMA

POSTED BY CHARLIE:

From the time of my second surgery on August 3, until a few days ago, I have been on narcotic painkillers and haven't trusted my brain. Also, I haven't been able to sit without pain for long enough to type out a blog entry. Moreover, as I have not been able to do much and have just started once again to see friends, there has been very little interesting in my life to report.

I feel as if I am crawling out of a miasma of pain, drugs, hospital admissions and readmissions, and, among other complications of the second surgery, an infection in the body cavity, a separate infection of the incision, and thirty-five pounds of edema in the lower body. Supposedly, there is about a four percent chance at MSKCC of getting an infection from one of the surgeries that I underwent. The doctors have no theories on why I got three infections from my two surgeries.

Since the second surgery, I have almost no appetite; most food does not taste to me as it always has, and the new tastes are unappealing. I am now allowed to drink wine, and I enjoyed a glass of rosé wine; but I couldn't drink the other two wines, a Champagne and a California cabernet, that I tasted. (My doctors say that, in time, food and drink will regain their tastes and appeal.) I can sleep for only about an hour before I wake up in discomfort.

I recall that Vince Lombardi was quoted to the effect that fatigue makes cowards of us all. He would have been right in my case. After a few weeks, it was hard to focus on the rationale that led me to agree to undergo these surgeries. I just wanted the pain and the unremitting feeling of sickness to stop.

Ideally, Dr. Saltz would like for me to be resuming chemotherapy about now; however, when he saw me six days ago he said that I had to gain weight by eating as much high-calorie food as possible, as well as gain strength through exercise, before they could start the chemotherapy—otherwise, a single dose might kill me. I weighed 161 pounds that day. This morning, I weighed 155, as a result of digestive tract problems that caused the doctors at MSKCC to send me to Urgent Care for most of the day. So far, the tests do not reveal any medical problem to be treated.

I have been freed to exercise again. I have started physical therapy, and I have reactivated my gym membership. I tire easily, but it feels great to be moving again and to observe some faint signs of rehabilitation. If you recall the model Twiggy from the 1960s, she was robust by comparison with me. I have lots of room for recovery.

I am feeling up to seeing friends again, if only for short visits. I hope to see you soon. With this update having been posted, we can talk about things other than my health when we get together. I think of the current period as a brief window of opportunity, as there is no way of knowing how sick the chemotherapy will make me.

SEPTEMBER 25, 2009
FRIDAY 8:36 AM

MISSY JONES

My cousin, Melissa Ellen "Missy" Jones, died at age sixty-one on September 5 in Virginia Beach. Although she had not been well, her passing was still a shock. Tragically, both of her siblings, Patsy and Cynthia, had died of unrelated causes as young adults, early in their married lives. Missy was the middle child and was unmarried. She left behind her nephew, Mundy Hackett, Patsy's son, a wildlife biologist and a good man; and her mother, Lucy, my mother's youngest sister and as fine a person as I have ever known. I can only wonder at Lucy's emotional balance, wisdom that enables her to find the good in life, strength of character, consideration of others, complete lack of self-centeredness, delightful personality, and general good nature; she exemplifies these virtues and more in the eyes of all who know her.

I have always felt close to the Jones branch of my family, and it was hard for me to accept that I was too ill to travel to Virginia Beach for Missy's Episcopal Church funeral service. No matter how difficult the circumstance, Missy had always made me feel especially welcome.

SEPTEMBER 28, 2009

MONDAY 10:00 AM

WHY ME?

From the time I was diagnosed with cancer, I have never thought, "Why me?" Shortly after being diagnosed, I can remember thinking, "Well, why not me?" For years, I have tried to be careful about diet, exercise, and checkups. So what? All that healthy living can do is improve the odds. Anyone can get cancer.

I must admit that when I think about my oncologist's estimate that someone with my cancer diagnosis who underwent the surgeries and chemotherapy slated for me should have about a 20 percent chance of being cured, I cannot help thinking, "Why me?" Why should I be the one out of five who makes it?

"UNFORTUNATELY, THERE IS BAD NEWS..."

"Unfortunately, there is bad news on your scan," Dr. Saltz told Susan and me during our visit to his office on Tuesday of last week. We were unprepared for definitive news, good or bad. The scan to which he referred was a CAT scan that had been taken the previous week for the primary purpose of establishing a baseline observation of my newly hypertrophied liver, the product of my most recent, August 3rd, liver surgery. The bad news was two apparent tumors in the liver and perhaps a tumor in a lung and another in my lower abdomen. Although it seemed obvious that Dr. Saltz was fairly sure about the tumors in the liver, he said that the next diagnostic step would be a PET scan, which I underwent two days later.

Today, we met again with Dr. Saltz, who reported to us that the PET scan "lit up" on three tumors in the liver, in addition to a region in the liver along the surgical line. That region might be more tumors or just inflammation. Although the spot in the lung also lit up, he is not sure whether or not it is a tumor. In any case, he does not seem to think the possible lung tumor is an important issue, given the tumors in the liver.

Dr. Saltz recommends that I move forward with a lighter program of chemotherapy than he had planned for me when the possibility of cure, however slight, justified heroic measures, such as the back-to-back surgeries I underwent this summer. I conveyed to Dr. Saltz that, with cure no longer a possibility, my objective in any treatment regimen would be quality of remaining days, not quantity of remaining days.

Dr. Saltz refused to be drawn into any speculation as to how long I may live. From the little that I've read and heard on the subject, I have had in my mind as a working hypothesis from the time of my diagnosis that a year or two might be a realistic life expectancy for someone in my situation.

I don't find myself thinking much about how long I may live. I do think a lot about what the rest of my life will be like and the decisions that I am likely to face along the way. I think a lot about the effect on my family of my remaining life, my manner of dying, and my being deceased.

Susan and I and our children have had a week to adjust to the roller coaster of going from no hope, to slight hope, to no hope of a cure. Now, I find myself turning my attention to the short-term objective of recovering sufficiently, through diet and exercise, from the surgeries and their complications to enable Susan and me to resume some semblance of a normal social life and perhaps even take a trip somewhere.

When I was a kid, my family spent our summers at Ponte Vedra, Florida. I love the ocean, and I am fantasizing about a sojourn at some posh beach resort. We'll see what happens. It seems that there is always something upon which to pin our hopes.

OCTOBER 7, 2009
WEDNESDAY 4:55 PM

SECOND GUESSES

Before I was diagnosed with cancer, it had long seemed to me, without looking into the matter deeply, that an all-too-common outcome for people diagnosed with late-stage cancer is that they undergo terribly debilitating treatments, and then die anyway. It seemed to me that if I ever got such a diagnosis, I should go to great lengths to avoid this trap.

My doubts about the wisdom of surgery, chemotherapy, and radiation in many such cases were reinforced by the experience of my Aunt Daisy, my father's younger sister. When Daisy was some twenty years older than I am, she was diagnosed with lung cancer. After considering the side effects of the potentially curative treatments, she decided to accept only palliative care. It seemed to me that she lived out the remainder of her life with great equanimity. In fact, when I would speak with her, she was always in good spirits and up-to-date on the latest developments in the outside world. She was even reading about nanotechnology, which she discussed with me, knowing that it was a professional interest of mine. As far as I know, she never second-guessed her decision to accept only palliative care.

Yet here I am, without hope of cure, physically shattered by two surgeries and their complications, getting ready to embark on a second, more protracted, round of chemotherapy. As someone who has always second-guessed his decisions, I have, as you might expect, been thinking about the wisdom of my decisions since my diagnosis.

The first decision, in early March, was to start chemotherapy rather than go into hospice care. That decision seemed easy enough. I was losing energy, weight, and muscle mass rapidly; wine had tasted strange for a few weeks, and food was starting to do so as well. If the chemotherapy became intolerable, I could always quit it and go into hospice care. In any event, although the chemotherapy produced a galaxy of unpleasant and sometimes frightening side effects, I regained energy and weight and got to spend a lot of time with family and friends over a three-month period (two months on chemotherapy, one off).

The second decision was to undergo colon and liver surgeries. After having been told from the time of inception that I was incurable, the unexpected chance at a cure was too precious a gift to pass up. It seemed to me that I had to undergo the surgeries, regardless of their certain trauma and potential complications. As my daughter, Elizabeth, pointed out to me recently, if I hadn't undergone the surgeries, I certainly would have second-guessed myself for the rest of my life.

How about the pending decision to resume chemotherapy? I think that with the particular cocktail the oncologist has in mind, there is only a small chance of something irreversible happening before I could stop the regimen. I'll probably go along with his recommendation. But I'm mindful of the pattern of my treatment decisions: It seems to describe the very trap that I always wanted to avoid.

OCTOBER 9, 2009
FRIDAY 6:19 PM

SAND IN MY SHOES

This weekend, although I am not taking the *Orange Blossom Special* to Florida, I am planning on getting "some sand in my shoes" and "losing those New York blues." Susan and I are accepting a long-standing invitation to stay with friends in their Long Island home on the oceanfront in Quoque. We will leave on Saturday morning and return to New York City on Monday morning. Traffic should be relatively light at those times: The ninety-mile drive should only take a couple of hours. Susan will drive. I will read the paper, if I don't snooze.

The weather forecast is favorable, and I am looking forward to opening the bedroom windows at night in our hosts' lovely second-floor guest room and sleeping and waking enveloped in the ocean's rhythms. When I was a kid, my architect father built first one, then, a few years later, another summer home for us on the oceanfront at Ponte Vedra, Florida. Neither had air conditioning; both were designed to maximize the circulation of the sea breezes. Although it got hot and humid some nights, a little discomfort was a low price to pay for the sounds and smells and breezes of the ocean.

This will be the second time since the end of February that I have left New York City. The first time was at the end of July, when

Susan and I went to Saratoga Springs, New York, for six days, just prior to my last surgery. The debilitating effects of the surgeries and their complications are waning. My taste buds are working again, and I have gained two or three pounds in the last few days. I am working assiduously at a rehabilitation facility and at a gym. I spend about two and a half hours a day on exercise and stretching. I am sleeping much better, sometimes for as long as three hours at a stretch, versus maximum sleeping stretches of an hour a couple of weeks ago. Nearly every day, I am again seeing friends, if only for short visits in our apartment.

The couple hosting us this weekend could not be warmer, more comfortable, or more fun to be around. The male half of the couple is also a New York Giants fan, so we will watch the Giants—the mighty, undefeated Giants—game on television on Sunday in the best of company. I wouldn't be surprised if we watched the Florida–Louisiana State battle of undefeated Southeastern Conference (SEC) rivals on Saturday night as well. My father is a University of Florida alumnus, so I grew up as a Fightin' Gators fan, back when the Gators were perpetual losers who had never won an SEC title, much less a national championship. Chomp, chomp.

OCTOBER 14, 2009
WEDNESDAY 4:33 PM

"THAT'S ALL"

According to Jane Brody in her *Guide to the Great Beyond* (Random House, 2009), Brooke Astor at age ninety wrote, "Death is nothing and life everything. That's all."

My appetite is improving enough so that Susan and I have gone twice with friends to good restaurants (once, as guests, to Daniel, which just was awarded a third star by Michelin) in the last few evenings and thoroughly enjoyed ourselves. In terms of stamina, strength, and flexibility, although I am still in quite diminished condition, I am making measurable progress during the two-hour workouts in which I engage most days. Friends who haven't seen me for a while are often kind enough to express how well they think I look—perhaps expectations have something to do with their reaction. Towards the end of this month, Susan and I are planning a brief holiday in Key Biscayne. In short, I am starting to be able to enjoy myself again and want to be as active as possible before likely resuming chemotherapy in early November. Even during this coming chemotherapy, based on my experience with most of the drugs that I am slated to be getting, I think it is reasonable to hope that I will have some relatively good days during each two-week cycle.

OCTOBER 16, 2009
FRIDAY 8:55 PM

FOUR SEASONS

Last night, Susan and I were honored to be amongst the guests at a surprise eightieth birthday party for a friend and mentor. As he and his wife and another couple entered the magnificent dining room—the Pool Room of the Four Seasons Restaurant, which had been reserved for the occasion—his assembled family and friends on the alcove above delighted in his genuine surprise and pleasure. In due course, making his way from friend to friend, he came to Susan and me. I got the sense that he might have been surprised to see me there, given my recent travails. We hugged, and he said, "If only I were God."

A MOVEABLE FEAST

*As I ate the oysters with their strong taste of the sea
and their metallic taste that the cold white wine washed away,
leaving only the sea taste and the succulent texture,
and as I drank the cold liquid from each shell
and washed it down with the crisp taste of the wine,
I lost the empty feeling and began to be happy and make plans.*

—Ernest Hemingway

ORANGE BLOSSOMS

Sorry for the silence. We enjoyed our holiday on Key Biscayne. If you haven't been to Key Biscayne, it is an island suburb of Miami, six miles south of Miami Beach, with a population of about ten thousand, reached by a four-mile causeway stretching over the Atlantic Ocean east of Miami. Some three decades ago, we stayed on Key Biscayne a number of times with our children, at a lovely, low-keyed resort, the Key Biscayne Hotel and Villas.

During the waning years of the heyday of Miami's Hialeah Park, when many of the powerful stables in the East and most of the horses prepping for the Kentucky Derby still wintered at Hialeah, we sent our more modest string of horses to Hialeah each winter as well, under the care of trainer LeRoy Jolley. During that era, LeRoy trained several champions, won two Kentucky Derbies (including one with the filly Genuine Risk for Diana and Bert Firestone), and became the youngest trainer at the time to ever be elected to racing's Hall of Fame.

Hialeah, with towering Royal Palms, wonderful tropical gardens, and spectacular flights of pink flamingos, was once Florida's leading tourist attraction. Eventually, Hialeah faded and closed. While it was still renowned as arguably the most beautiful racetrack in the world, we frequently visited our horses and went racing there in the winter. On Sunday mornings, LeRoy and I, and often Susan, would fly in the dark on a rented helicopter roughly one hundred miles north, to what was then called the St. Lucie Training Center, to watch the *babies* (the yearlings or two-year-olds, depending on whether it was before or after New Year's Day) work. In LeRoy's program, the riders worked the babies in fast times. Nowadays, thoroughbreds seem more fragile, and trainers don't push young horses. Even by the standards of three decades ago, LeRoy was aggressive. He would pair the babies off with the winners from last week matched against each other. After a few Sundays, you had a pretty good idea of what you had.

Each winter, we still send horses of various ages to that same training center in the winter. It is now called *Payson Park*. The training center is much better maintained but lacks the occasional thrills, in the form of rattlesnakes and alligators, that it offered as the *St. Lucie Training Center*.

Flying back to Miami, in the daylight, we might detour to fly along the Atlantic coast from Palm Beach south, looking at the mansions along the shore and at the swimmers in the ocean, who were blissfully unaware of sharks circling nearby. The pilot was the son of a horse veterinarian; he had flown helicopters in Vietnam and been a killer-whale trainer at the Miami Seaquarium. (He had quit his job there after a male orca, which later turned out to have a fatal brain tumor, grabbed his head and held him underwater until, in desperation, he punched it in the eye.) One day I talked him into buzzing the pool twice at the Key Biscayne Hotel and Villas when I saw that Susan, Elizabeth, and David were sunning themselves there. (They didn't even look up.) Key Biscayne was still undeveloped enough to permit the pilot to hover over a vacant lot across the street from the hotel and let me hop out. (Someone got the number of the helicopter, and the pilot got into trouble for that little stunt.) We stopped going to Key Biscayne when the hotel's oceanfront property was acquired for replacement by a much grander Ritz-Carlton resort.

We were delighted to find that one of Key Biscayne's two large public parks, Crandon Park, which occupies the entire northern end of the island, is still the attractive playground that we remembered. Crandon Park has ample free parking spaces and features not only extensive beaches but also playing fields, picnicking areas, an eighteen-hole golf course, and tennis facilities that host a professional tournament each year.

The best surprise for us was the Bill Baggs Cape Florida State Park at the southern end of the island. Not only is its early nineteenth-century lighthouse now meticulously maintained, thanks to a private foundation, but also the park service was able to replant the park in indigenous vegetation after Hurricane Andrew denuded the island in 1992 of the hardy, non-native, Australian pines that for years had cut off sunlight and pulled the water out of the thin soil. Andrew, one of three Category Five hurricanes to come ashore in the United States in the twentieth century, struck the mainland south of Miami. Key Biscayne, always the first area of Miami to be evacuated when a hurricane approaches, was hit hard. But for the Bill Baggs State Park, "It's an ill wind that blows no good."

NOVEMBER 1, 2009
SUNDAY 5:58 PM

NOT THE NEW YORK MARATHON

The New York Marathon had its fortieth running today, in ideal weather for runners. For the first time since 1982, an American won.

We won a less important race today—the ninth race at Aqueduct Racetrack, at a mile and a sixteenth on the grass for horses bred in New York that had never won a race. Our horse, a three-year-old gelding named Mustang Island, was making his first start on the grass and the second start of his career. Because of a minor injury, he had not run in nearly a year and was too keen in the early stages of the race, fighting with the rider, Joe Bravo, to be allowed to extend himself. Once Bravo swung Mustang Island out in the stretch for running room and let him go, Mustang Island swept by the pacesetter and won going away.

Not only was the race *not* the New York Marathon, it also did not consist of a particularly good field of horses. But it is always fun to win a race and good to collect a purse. (Seven-to-one betting odds added a little spice as well.) Whenever one wins with a horse that has been asked to do something for the first time (in this case, run on the grass), and it accomplishes what one has asked of it rather easily, one cannot help dreaming a little.

NOVEMBER 7, 2009

SATURDAY 7:45 PM

THE BREEDERS' CUP CLASSIC

Thoroughbred horse racing's most important annual two days of racing—in the United States and arguably the world—concluded today. The richest of those events, the $5 million Breeders' Cup Classic, was won for the first time by a filly or mare, Zenyatta, which closed out her undefeated career with her fourteenth victory. The second-place finisher to the favored Zenyatta was long shot Gio Ponti, trained by Christophe Clement.

Since 1995, Christophe has trained most of our horses. We feel close to him and his wife, Valerie, and their two children, Miguel and Charlotte, and we are proud of his well-deserved, growing success. When we first sent our horses to him, he was still in his twenties and early in his career as an independent trainer. Friends of ours had had horses in France with his late father, the successful trainer, Miguel Clement; Christophe's brother, Nicolas, had visited us at Saratoga Springs as a teenager. I had followed with interest Christophe's debut as an American trainer.

When I decided to switch trainers, I asked Christophe to take our horses, on the theory that if he was as good a horseman as I thought he was, he would soon be too popular to have room for our horses unless we were an established client. Once again, as happened to me so often over the years, I got lucky by giving a younger person a chance before it was obvious that he was ready.

It has been a long while since Christophe has needed our horses in his stable. In an article in the *New York Times* two days ago, Joe Drape noted that of this year's top ten American money-winning trainers, Christophe is the only one who has never had a horse in his care that tested positive for a prohibited medication violation.

Through talent, hard work, and superior organization, Christophe has been able to win important races while conceding an edge to those less scrupulous. Although he has quietly gone about his business rather than be strident or self-righteous about the improper use of drugs, he has shown courage on the topic when asked to do so. In 1999, I wrote an article, entitled "Full Disclosure," for the October 23rd issue of a leading trade publication, *The Blood-Horse*, about medication practices in the industry in general, and in the state of Kentucky in particular. To corroborate what I had to say, I asked Christophe if I could cite something in the article that he had told me. Unlike me, a minor owner/breeder with nothing to fear from those who would resent such exposure of industry practices, a young trainer like Christophe had a great deal to lose by gratuitously making enemies. Nevertheless, he assented to my use of his name in the article.

Before publishing the article, the then-editor of *The Blood-Horse*, Ray Paulick, who certainly had to be concerned about industry backlash, asked me what I hoped to accomplish by writing it. I replied that I hoped to bring hidden practices into the open and start a dialog. (Of course, I was also angry and frustrated. As Susan, says, "It wouldn't be fun to win by cheating, but it's no fun being beaten by cheaters.") To my surprise, the article kicked off a furor, including articles by others. In Kentucky, a horse-racing medication law was passed and a leading state racing official lost his job. If Christophe's association with my article caused him any problems, he never complained.

LAST TANGO?

One week ago, I underwent a CAT scan, and on Thursday of last week, I had an appointment with Dr. Saltz, who indicated that "everything is growing, but not rapidly." In tacit keeping with my previously stated objective—now that the possibility of cure is off the table—of maximizing the quality of my remaining days, Dr. Saltz gave me the option of postponing for a while longer the resumption of chemotherapy. Two thoughts flashed through my mind: I could fly down to Jacksonville, Florida, to visit my parents in nearby Ponte Vedra; and I could go with Susan to Paris, getting back to New York in time for our daughter Elizabeth's visit to us from November 20 through the Thanksgiving holiday.

For months, I have wondered if I would ever see my parents again, and I have been thinking it unlikely that I could go abroad again. Susan was with me last Thursday in Dr. Saltz's office, and I asked her if she would prefer a place other than Paris. She loves Paris and was not hesitant in her response.

I just got back from visiting my parents in Ponte Vedra. Tomorrow, I will see Dr. Jarnagin, the surgeon who operated on my liver. Wednesday, Susan and I will depart for Paris. Although I have some physical problems caused by the cancer and by the surgeries, and their complications, I feel better than I have since February, and I may never again feel as good as I do now. Importantly, my taste buds are once again in good shape—a prerequisite for a sojourn in Paris.

LIVING WELL IS THE BEST REVENGE

WHITE TRUFFLES AND THE HAND OF GAUL

In France, November is in the season for wild game, oysters, mushrooms, and white truffles. This year, it was also the time when Algeria played Egypt and France played Ireland for slots in next year's World Cup *football* (soccer) finals.

Although Parisians seem to refer to "The Crisis" ("La Crise" in French) in the past tense, it is still much easier to get a table in most good restaurants than it was a couple of years ago. As you would expect, at most of the restaurants where we dined, the chefs incorporated seasonal delicacies in their menus. Most memorable to me are the white truffles, the mushrooms (especially the cepes), the oysters, and the grilled, stuffed pigs' feet.

Both yesterday and on the second day of our stay, we ate lunch at Huitrerie Regis, which is located on the Left Bank, near the covered food market just off the Boulevard Saint-Germain. This wholesaler of oysters to other Parisian restaurants also has a simple restaurant that can barely hold its seven tiny tables, each with two seats. The menu consists primarily of the oysters that have arrived that day; the wine list has a Muscadet and several Sancerres.

On one of the days when we ate lunch there last year, the menu also offered sea urchins; I have not had a chance to eat sea urchins often, but they were by far the best that I have had. There were bright orange, flavorful, wild shrimp on the menu on both occasions; and yesterday, there were also Belon oysters—the big ones, Belons No. 00, not the smaller oysters, Fines de Claire and the Spéciales de Claire. For some thirty years, I have carried the memory of Belons No. 00 that I ate at Prunier, the famous seafood restaurant on Avenue Victor Hugo in Paris. The Belons that I ate at Huitrerie Regis yesterday matched those of that memory.

I didn't realize until this week that sports fans in the United States are not serious. Algerian soccer fans are serious. Algeria played Egypt on Wednesday in Khartoum, Sudan, after having lost to Egypt in Cairo last Saturday. Before Saturday's game, Egyptian fans had stoned a bus carrying the Algerian players, inflicting bloody head wounds on three of the rival players. In the ensuing riots, thirty-two people were injured, and Egyptian-owned businesses in Algeria were attacked. For the rematch in Khartoum on Wednesday, the Egyptian and Algerian fans were housed miles apart, with fifteen thousand riot police patrolling the streets.

I had no idea that so many Algerians live in and around Paris. The streets of Paris were mobbed with cars and pedestrians waving Algerian flags, following Algeria's victory on Wednesday. We were caught in a taxi during the melee. Although our taxi was hit on the windshield by a wet projectile, we were merely inconvenienced and never felt in danger. (Why should there be any danger? Algeria had won.)

Later Wednesday night in Paris itself, France was playing Ireland. Ireland, which had not won a match in France since 1938, was leading 1 to 0 until late in the game, when a French player's hand touched the ball in passing it to a teammate, who then scored. The referee missed the impermissible hand foul, and the tie put France into the World Cup finals. As we tried to fall asleep in our hotel room, the raucous Algerian fans were joined by the more decorous French fans in the streets of Paris. European newspapers yesterday were filled with coverage of the outrageous result of France's game with Ireland. They referred frequently to a similar situation in the 1986 World Cup, just four years after the Falkland Islands War. Argentina's Diego Maradona won the match by punching the ball into England's goal. Maradona referred to his bit of cheating as "The Hand of God." London's *Daily Telegraph* dubbed this French incarnation "the Hand of Gaul."

NOVEMBER 25, 2009
WEDNESDAY 5:32 PM

THANKSGIVING

I have much for which to be thankful this year. Everyone else in my family is in good health. Our two children are here with us for the holiday. (Today, we went to a Broadway play, *A Steady Rain*—two fine actors, Hugh Jackman and Daniel Craig, in a slight play.) Three close friends will be joining us at the Thanksgiving table. As usual, all are eager to do justice to Susan's annual feast. As I write this entry, my appetite is being tantalized by smells from the kitchen. I am grateful for the respite that I have been allowed from chemotherapy and expect to be able to hold my own at the dinner table. Indeed, I feel better than I have since early this year, a month or so before my cancer diagnosis at the beginning of March.

The poignancy of knowing that it is not unlikely that this Thanksgiving will either be my last or, if not my last, the last that I will be in any condition to enjoy, heightens my appreciation of the occasion. During each of the few trips that I have taken since March—to Saratoga Springs, Miami, Jacksonville, and Paris—I have been constantly cognizant that I might be seeing a familiar place for the last time. Whenever I have seen a friend who lives somewhere other than New York City, I have wondered if it were for the last time—a thought that I assume has been reciprocated.

DECEMBER 3, 2009
THURSDAY 5:22 PM

INTERLUDE

This afternoon, I had an appointment, scheduled about a month ago, with Dr. Saltz. The purpose of the appointment was for him to determine whether, given my objective of maximizing the quality of my remaining days, I could continue my respite from chemotherapy through the holidays. Because I am feeling so well, I was pleased but not surprised when he agreed that, unless I have a turn for the worse, I can extend my holiday into January.

I realize that I may never again be as free of pain, or as energetic, or as able to taste food and otherwise enjoy life as I am now. During this interlude between the miasma of the surgeries and the upcoming poisoning of more chemotherapy, I hope that I can keep the perspective of finiteness in the forefront of my mind from moment to moment, in the precious minutes, hours, and days ahead.

During this interlude, my major time sink is spending a couple of hours a day in either physical therapy or at the gym. I think that this expenditure of irreplaceable time is worthwhile. I am given to understand that patients with advanced cancer not only feel better, but also may live significantly longer if they follow a program of vigorous daily exercise. In my situation, I know of nothing else that does not require a trade-off of quality of life for quantity of life. The exercise has helped me to rebound physically from the two surgeries and their complications.

I experience a considerable psychological boost from doing something active to fight back. While I know that the cancer is winning and will kill me, struggling against it feels good. If, today versus yesterday, I can lift more weight, stretch over a wider range, and go faster, longer on the cardiovascular machines, am I dying? When I have added more weight to the bench press and am completing the last of the repetitions that I have assigned myself, as I did today: In that moment, I have won. I have won.

During the more broadly defined interlude between diagnosis and death, the time-consuming writing of this entry is, for me, another way of struggling, of fighting back.

DECEMBER 8, 2009
TUESDAY 4:34 PM

PAINFUL LIMITS

When I go back on chemotherapy, I will probably receive Avastin in conjunction with the same cocktail of poisons, FOLFIRI, that I received when I was on chemotherapy previously. Britain's National Health Service will not pay for Avastin even though, in conjunction with chemotherapy, Avastin shrinks tumors in seventy-eight percent of patients with my condition. That benefit sounds impressive until one considers that it translates into median survival of 21.3 months, as opposed to 19.9 months without Avastin. A modest improvement in median survival does not, of course, mean that some patients do not benefit significantly from Avastin.

Given our society's many needs and the budget deficit, should Medicare pay, as it will, for my Avastin? Although I have not attempted to follow all of the twists and turns of the healthcare debate, I do not think that the cost-containment proposals currently under discussion by our politicians in Washington include consideration of eliminating Medicare payments for things like Avastin (especially, one could argue, for folks who could afford to pay for it themselves). Until our nation has completed the squandering of the U.S. dollar's status as the world's reserve currency, I wonder if our politicians will have the temerity to suggest to us, the American people, that we will have to accept that there are painful limits to what our country can afford, whether it be for medical care, other entitlements, bailouts of businesses, pork-laden stimulus programs, or some notions of national defense.

DECEMBER 14, 2009
MONDAY 4:09 PM

HALLELUJAH

From August 22 through September 12, 1741, George Frideric Handel composed in London a rough score of *Messiah*, consisting of some 250,000 notes written with quill pens. (Presumably, he wasn't spending much of his time during those twenty-four days watching television or on the Internet.) One of the earliest performances of parts of *Messiah* in the New World took place in October 1770 at Trinity Church, at the same location at the foot of Wall Street currently occupied by Trinity in its second reconstruction, consecrated in 1846.

Handel conceived of *Messiah* as an oratorio to be performed during Lent, just prior to Easter. In recent years, performances of this retelling of the life of Christ have become part of the Christmas tradition. This year, two performances were scheduled at Trinity Church: one for yesterday afternoon, which Susan and I attended; another for tonight. I doubt that there was an empty space in a pew or an unoccupied folding chair in the aisle (where we sat, near the front) of this beautifully proportioned, perfectly maintained, Neo-Gothic church.

To my tone-deaf, untutored ears and to my dim vision, yesterday's performance of my favorite Christmas music was completely satisfying—expertly and charmingly conducted and performed, in an appropriate and sublime space.

My Christmas season, my favorite time of year, had begun. Hallelujah.

DECEMBER 19, 2009
SATURDAY 10:04 PM

A MOVEABLE FEAST

Since we arrived back in New York from Paris about a month ago, I have taken advantage of my unscheduled interlude by eating lunch with one or more friends almost every day and attending some sort of party or dinner with Susan and one or more couples almost every night. The only ill effect to report from all of this activity is that I had to take all of my trousers to the tailor to have them let out an inch.

I have continued to go either to physical therapy or to the gym almost every morning, and I have steadily regained energy, strength, and flexibility. Although the surgeries and their complications have left their marks, I am in good spirits, eating and drinking con brio, and feeling remarkably well as I look forward to making plans to spend time in the New Year with other family and friends.

On Monday morning, Susan and I will head south by auto-mobile. We will divide the driving in our usual fashion. She will drive, and I will read. She can't read in a moving vehicle, and I can't stand to waste time. If the snowplows do their job, we should arrive on December 23rd in Ponte Vedra, Florida, where David will join us for a Christmas visit to my parents. The next morning, the three of us will drive west from Ponte Vedra across the state to St. Petersburg, to spend Christmas Eve and Christmas Day with Susan's father and his friend, Danka, and one of her brothers, Tom, his wife, Judith, and their son, Nick.

On the 26th, Susan and I will travel to the middle of the state to Ocala, to visit friends and horses, including Statesmanship, the most successful horse that we have, to date, ever raced (in part-nership with Mike Rankowitz and Peter Karches). Thanks to the kindness of friends, Jill and John Stephens, Statesmanship is now a happy Florida retiree.

We will wend our way up the Atlantic coastline, visiting peo-ple who cannot easily travel—a friend in Charleston, my Aunt Lucy, and an unrelated couple in Virginia Beach. We intend to arrive back in New York City on New Year's Eve.

Before I retired at the end of last year, Susan and I had intended to take more advantage of New York than I had ever had time to do while working. We had envisioned spending leisurely lunches at fine restaurants before heading off to museums.

My extended interlude should enable us to realize a bit of that retirement dream. During the first week of January—a quiet week in New York, free of crowds—we are going to pretend that we are foreign tourists in New York. Today, I made luncheon reserva-tions at Bouley, Aquavit, Le Bernardin, Jean Georges, and Eleven Madison Park. Tomorrow, we will pick concerts and plays. We won't decide now on museums: I can't think of many more pleas-ant topics to save for luncheon conversations than discussions of which exhibits to visit that afternoon.

2,765 MILES

Susan and I made it home last night, on New Year's Eve. During our mini-odyssey, she had driven 2,765 miles and was too tired to go out to dinner, let alone cook. We ordered take-out food from our favorite local Chinese restaurant, Grand Sichuan. We washed the spicy food down with sparkling Gewurtztraminer from Navarro Vineyards in California. We did not miss going out and toasting the New Year with Champagne.

The trip was a great success. We had a wonderful Christmas, highlighted by a delicious Polish Christmas Eve dinner cooked and hosted by my father-in-law's friend, Danka. Along our route, we saw as many as possible of our friends and family who cannot easily travel. Two originally unplanned visits fell conveniently into place. We saw my cousin, Jo Erickson, née Lane, at her beach house along our route in North Carolina; and a boyhood friend, George Martin, and his wife, Pat (who makes the best pecan pie that I can recall), were kind enough to drive four hours from their home to have dinner with us in Charleston.

Nothing is ever perfect. I wish that we had had time to detour to see even more friends and family; I wasn't happy about not having time for exercise for that many days; and the food at most of the restaurants along the way goes far towards explaining why there is an epidemic of type 2 diabetes in this country. At Catherine's Restaurant in Ahoskie, North Carolina, where they served butter with the hush puppies, the featured dessert—fried cheesecake—might have been one of the healthiest choices on the menu.

WHAT GOES THROUGH MY MIND

On our recent trip to the South, a friend in Charleston—an artist—asked me, as nearly as I can recall her words, "What goes through your mind since you were diagnosed?" Although my thinking about how to handle dying of cancer continues to evolve, one constant is that I have never once thought, "Why me?" Although I am cognizant of ironies in my case, such as having been a bit of a health nut and having extremely long-lived parents, I have consistently thought, "Well, why not me?"

For the first couple of months after I was diagnosed, whenever I awoke, the thought shot through my mind, "I am dying of cancer." For the next few months—during the period when it was thought by my doctors that surgeries combined with further chemotherapy would give me about a twenty percent chance of becoming cured—whenever I awoke, the thought shot through my mind, "I am probably going to die of my cancer, but there is a chance of being cured." After tumors reappeared in my liver soon after my second surgery, the thought that shot through my mind upon awakening reverted to, "I am dying of cancer."

For the last couple of months, no particular thought is consistently in my mind as I awaken. Although I do not recall my dreams, I do not have the impression that they have been nightmares or other than random. Currently, whenever I am conscious, I simply have a chronic background awareness that I am dying of cancer.

Upon learning that I had incurable cancer, one of my immediate thoughts that has not been modified by time was that such a diagnosis was inherently isolating and that I must fight against isolation. Another thought that came to me shortly after being diagnosed and that has not been modified by time was that I must not allow myself to become depressed. Depression, I thought and think, is a luxury that I cannot afford: I don't have much time; if I were to allow myself to become depressed, it would only exacerbate my physical suffering; and if I were to allow myself to become depressed, it would only add to the burden of those around me, especially my family. A corollary realization of mine that has not been modified by time is that one's responsibilities to others are not diminished by fatal illness—they are increased.

I dread further treatments for cancer—being poisoned by chemotherapy, butchered by surgery, and beset by hospital-incurred infections and other complications of cancer treatments. I dread the repeated trips to, and long waits in, the depressing, overcrowded, understaffed Urgent Care Center. I dread the helplessness, monotony, and debilitation of being confined to a hospital bed again.

Except for my concern for the impact on my family of my permanent absence, I have rarely found myself thinking about being dead. I do dwell on being drugged and physical suffering and loss of consciousness and dignity while dying of colorectal cancer. I do dwell on what I might or might not be able to do, taking into account my family's sensibilities, to exert some control over the process of dying. Since long before I was diagnosed with cancer, I have thought that perhaps the saddest words that are commonly uttered upon the death of a friend or loved one are, "It was a blessing."

JANUARY 7, 2010
THURSDAY 6:14 PM

THROUGH TOURISTS' EYES

Susan and I had planned to spend from this past Saturday, January 2, through tonight trying to see Manhattan through tourists' eyes. Having recently refreshed as tourists our memories of Miami, Paris, and the heart of the eastern seaboard of the United States, including

Savannah and Charleston, we felt ready for our job of touring Manhattan. During this annually quiet week in Manhattan, we knew that it would be easy to make reservations and hail taxis.

Since Saturday, we have eaten lunch at Bouley, Aquavit, Le Bernardin, Jean Georges, a private club, and Eleven Madison Park. Leaving aside the private club (at which we were guests and had a first-rate meal in august surroundings and delightful company), each of these restaurants was as excellent in its own way as we had recalled; as a group, they were, we thought, competitive with the top restaurants in Paris. Moreover, all of these restaurants offer real value for money at lunchtime, as compared with their prices at dinnertime. Nowadays, hearing few American accents in such New York restaurants at midday, it is easy for an American diner to imagine that he or she is a foreign tourist. For an actual foreigner paying in depreciated American dollars, these world-class restaurants are an incomparable bargain, even for dinner.

In the evenings, we went to a revival of *The Emperor Jones*, to the Metropolitan Opera for Puccini's *Turandot*, to off-Broadway for *Flaminco* by Soledad Barrio and Noche Flamenca, and to Broadway for the musical *Jersey Boys*. David joined us for *Jersey Boys*, as he did tonight for a preview of a revival of Arthur Miller's *A View from the Bridge*, with Scarlett Johansson, who was eclipsed in this preview of her Broadway debut by her co-star, Liev Schreiber. Tomorrow, Susan will go to the West Coast for a few days, I will have lunch with friends at Asiate, which is in the Mandarin Oriental Hotel, and tomorrow night David and I will go to Madison Square Garden to see the bull riders in action.

Even though we barely had time in these past few days to glance at a few of Manhattan's cultural attractions, our glimpse through tourists' eyes reminded us of why New York City was the U.S. city most visited by tourists last year. It reminded us of why we raised our family here, and why we made our careers here. It reminded us of why we had planned to continue to live here after I retired.

From experience, I know that once I resume chemotherapy, food and wine may become unappetizing, and it will be difficult for me to stray far from MSKCC's facilities. If my next course (which might last six months or so) of chemotherapy goes well, I should recover from it sufficiently for my sense of taste to return and to allow us to have another period of travel. Whatever happens, we are happy to be tourists-in-residence in New York, where we are blessed with friends, and where the only non-monetary limiting factors on things to do and great places to eat are one's interests, curiosity, time, energy, and appetite.

01 11 10

Happy Palindrome Day. On Wednesday, Susan and I will be off to Jupiter, Florida. South Florida's record cold weather is forecast to give way that day to a more seasonable high of sixty-five degrees. We will stay on the oceanfront and visit some friends and horses. By now, the horses in Florida must all have grown their winter coats, and the snowbirds must wish that they had migrated South with their winter coats.

Before we return to New York late on January 20, I want to play some golf. It has been almost a year since I last played, and, until recently, I thought that I could never play again. Nonetheless, we have kept our membership in a golf club in Tequesta, Florida, and I am getting stronger and more flexible every day. Why not give golf a try? Not that I can expect mercy from Susan on the golf course. Once I hit my ball in a bunker near an alligator, and Susan made me play it as it lay. She has always maintained that the alligator wasn't big enough to warrant a free drop and that, in any case, she was keeping an eye on it for me while I addressed the ball.

A GOOD MAN

If a man plants a tree, he knows that other hands than his
will gather the fruit; and when he plants it, he thinks
quite as much of those other hands as of his own.

—Alexander Smith
DREAMTHORP

A GOOD MAN

My father died this past Sunday, January 17, at approximately 5:25 p.m., in Jacksonville, Florida. He was born there also, on April 4, 1910. Tomorrow, he and my mother would have celebrated, quietly, their seventy-first wedding anniversary.

My father, Clyde E. Harris, went by the name Clyde. I am named after his father, Charles Eugene Harris, who grew up on a farm in Harris County, Georgia, and was vigorous until he died in his sleep at age ninety-one. Dad's mother, Daisy (née Hayes), was from Fernandina Beach, Florida. Judging from the one photograph of her that I have seen, she was an attractive brunette.

Dad spoke rarely about his childhood. He had a younger sister, also named Daisy. We inferred enough, and learned enough from Daisy, to know that, after their mother died in the 1918 influenza pandemic, the two children were shunted between relatives. Dad had also been infected by that "Spanish Flu," and this ability to recover from it may have been a hint of his robust constitution and long life.

Encouraged by a high-school teacher in a drafting class, Dad determined that he would become an architect. In 1932, upon graduation at or near—I'm not sure which—the top of his class from the architecture school at the University of Florida, he apprenticed in the office of Jacksonville's leading architect, Mellen C. Greeley. It was the nadir of the Depression, and there had been overbuilding in Florida during the land boom prior to the great hurricane of 1926. There was little work for architects, and no pay for an apprentice. Dad lived by borrowing five dollars a week from his father and occasionally getting freelance work.

With some revival in building activity in the mid-1930s, Dad's talent and outgoing personality won him projects, and he opened his own firm in Jacksonville in 1935. During World War II, he closed his office and worked on construction of military facilities, first for the Jacksonville office of the Army Corps of Engineers and then for his father-in-law's civil engineering firm in North Carolina. After the war, he reopened his office and practiced architecture until failing eyesight caused him to retire in 1997. For a few years, my older brother, Bill, joined him in his practice. According to billing records, operating a firm of only six to eight people, Dad designed at least eleven hundred projects, from sizable buildings to small renovations. Including his work before World War II, he may have executed as many as fourteen hundred commissions. Although he was best known for his designs of high-end residences and his creative remodeling jobs, he also designed schools, houses of worship, country clubs, post offices, small office buildings, and at least one high-rise apartment building and one manufacturing plant.

Dad loved practicing architecture. Towards the end of his career, he raised his fees and limited his practice to a few high-end residences a year. He thoroughly enjoyed practicing on this limited basis, which kept him in contact with younger members of the Jacksonville community. Once, while he was still working, he told me that he had no interest in retiring, as he saw no advantage in it: "I want to play golf only three days a week, and I do that now."

Dad liked swimming in the surf, surfcasting, and power boating. He liked deep-sea fishing and indulged my passion for that sport by taking our boat, under the expert guidance of Chuck Webb, our Ponte Vedra neighbor, out through the mouth of the St. Johns River and into the Atlantic Ocean. (The St. Johns is the *raison d'être* for Jacksonville and one of the many major rivers in the world, along with the Amazon and the Nile, that flows north, away from the Equator.) But Dad's sporting passion was for golf.

Dad caddied as a boy, and he began playing golf during World War II. His swing was unorthodox—he would rise up, then smash down on the ball—adapted from his years of playing handball. Heavyset, with powerful arms and legs, he became a low handicapper, smiting full shots long distances and using his deft architect's hands to advantage around the greens.

Dad liked playing golf on the links course at Ponte Vedra. He was happy playing with his pals at his home course, the Donald Ross–designed Timuquana. Dad had won the assignment of redesigning the clubhouse at the Timuquana Golf Club, and he was particularly proud of the result. At age fifty, he was senior golf champion at Timuquana, and on his seventy-second birthday, he played there in an effort to shoot his age: He birdied the eighteenth hole for a round of seventy-one, one under par.

Dad was a great guy. High spirited in his younger years, he was popular and always had a wonderful sense of humor. His father had had an endless supply of off-color jokes on any topic. Dad's favorite jokes tended to be more along the lines of political incorrectness. His quick, droll wit punctured pretense without giving offense. In his final years, no longer able to play golf, he repeated himself more than he had recounting tales of his golfing exploits, some of which, while true, were as unusual as his golf swing.

Although Dad had little interest in buying clothes or personal items other than golf clubs for himself, he enjoyed buying nice things for my mother. He also bought furnishings and paintings for our homes. Beige leather boating shoes with crepe soles were his preferred footwear with a tuxedo.

Dad did like stylish cars, and he owned a number of fancy automobiles over the years, including a Graham-Paige; a Brewster-bodied Jaguar Cabriolet; a four-door, pale yellow, Lincoln Continental convertible, with black leather upholstery and a white top; and an emerald green Chrysler Imperial with gunsight tail lights and camel-hair upholstery. He purchased cars rather impulsively. A test drive consisted of ascertaining whether the car had enough headroom to allow him to wear his hat while driving. Once he cut off the top and bottom of the steering wheel of one of his cars because he fancied the look of the controls in airplane cockpits.

Stoic and unselfish, Dad was a kind person who was unfailingly courteous to those less fortunate and generous to those in need. Quietly conservative in his politics, he was a practicing, believing Episcopalian, without religiosity. He spoke in complete sentences, enunciating carefully, at a measured pace, in his deep voice. He wrote well, in clear, economical prose. Mother says, "He never complained. He let me do that."

Little children delighted Dad, especially when they were being a little naughty. During my childhood, the prevailing child-rearing wisdom was, "Spare the rod and spoil the child." For him, that philosophy applied to unruly boys. As the youngest child and the only girl, my sister, Katie, was the apple of his eye, and he unabashedly tried to spoil her with affection and presents. Though not begrudging of praise for his children, he expected people to make the most of their opportunities, and he was never insincere. His rare compliments meant a lot.

Towards the end of his life, physical frailties were taking their toll. He had never heeded notions about healthy living. But his mind and sense of humor remained sharp, and he remained interested in everything from sports to domestic politics and world affairs. His repetition of favorite exploits was no indication of senility. He had retold stories for years. He simply couldn't resist sharing his joy.

Last Friday, at about 3:00 p.m., Dad got up to change a disc from which he and my mother were listening to the reading of a book. Before he could make it back to his chair, he fell backwards and broke his hip. As I am still on leave from chemotherapy, I was fortunate enough to be able to be at his bedside by Saturday afternoon. By then, he was on heavy pain medication, and the doctors had ascertained that he had pneumonia, which he probably had endured, without complaint, for some time. A combination of atrial fibrillation and the pneumonia had prevented the surgeon from repairing the severe fracture before I arrived.

Facing death, through a fog of pain, fatigue, and medication, Dad remained stoic and calm. He was characteristically cooperative, well spoken, and courteous towards the staff as they performed their procedures on him. He smiled in response to my feeble attempts at humor.

The attending physician asked Dad if he needed more pain medication for his hip. Dad replied, "It's okay." The physician said to me, "With him, everything's always okay." The physician then said to Dad, "You're a tough guy, aren't you?" Dad smiled slightly. He was dying as he had lived.

Saturday night, Dad accepted wearing a mask, to aid in breathing, that many patients cannot tolerate, and by Sunday morning he had rallied enough for the surgery to proceed. As he was wheeled away to the operating room, his grip on my hand was as warm and powerful as ever, and he was still smiling when appropriate.

While we waited for Dad to be returned to the Intensive Care Unit, Mother noted that it was perfect golf weather, cool and sunny. The surgeon came to us and told us that Dad had done well through the surgery, and Dad was subsequently returned to the Intensive Care Unit. He never regained consciousness. One of the female nurses, amidst their and our tears, asked Mother the secret to having a seventy-one-year marriage. Mother paused briefly. Then she replied, "A good man."

ROCK ON

Today, I had a regularly scheduled appointment with Dr. Saltz, who told me that my tumors have continued to grow, with the ones in the liver continuing to be more advanced than those in the lungs; and that they are definitely large enough to make a case for putting me back on chemotherapy now. Because my tumors were responsive to first-line chemotherapy last year, he thinks it is quite likely that he could make my CAT scans "look prettier" with chemotherapy. Given my priorities and that I feel so good now, there is also a case to be made for more "watchful waiting" before resuming chemotherapy.

For me, it is an easy decision: a choice that will, among other things, allow me to attend my father's memorial service on February 15. Then, Susan and I are likely to travel abroad.

GROWING UNEASE

As my regularly scheduled appointments for cancer tests and examinations approach, I often reflect that it must be much easier for me as an incurable to maintain a reasonable degree of equanimity than it would be if there were any hope of my surviving over the long term, much less being cured. After all, as I already know that I am going to die from my cancer, what really is there for me to fear in a new test result? I think of the apprehension in advance of receiving such test results others must suffer who have everything to lose—those who have been diagnosed with less advanced stages of cancer or who have been treated with apparent success. In terms of dread, is not my path easier than theirs?

Nevertheless, starting a week or so before I am to receive a cancer test result, I find myself gnawed by growing unease. Although this unease is not as great as it was in advance of my bimonthly glaucoma tests, it is enough to unsettle me.

I find that my fear of the potential horrors of the process of dying has been pushed into the recesses of my mind by my dread of resumption of chemotherapy. I have no doubt that dying, in its turn, will loom larger than anything I can now imagine.

FEBRUARY 6, 2010
SATURDAY 4:55 PM

SIXTY-SEVEN YEARS

Today, with gratitude and mild bemusement, I greet my sixty-seventh birthday. More than once in the last eleven months, it crossed my mind that I might not live to see it. Daddy Bill died of heart failure when he was sixty-seven.

Knowledge that I have terminal cancer has heightened my consciousness of the irretrievability of each second ticking away. In my adolescence, the poignancy of time passing was at least equally intense. Clocks, watches, and sundials have always fascinated me. But neither time-keeping devices, nor the workings of memory, can recapture youth's ferocity.

MEMORIAL SERVICE FOR MY FATHER

On Monday, February 15, the memorial service for my father will be held at Christ Church in Ponte Vedra. Tomorrow, Susan and I will fly down to Jupiter, Florida, to enjoy the balmy weather, play golf, see horses, and stroll on the beach. Later in the week, we will drive north to Ponte Vedra to join my gathering family.

POSTCARD FROM THE SEASHORE

As I type this from our hotel room, just above swaying, rustling palms and about forty yards from pounding surf, news reports of near-record snowfalls in the Northeast are enhancing my appreciation of the blue sky and bright sunshine outside. We are staying at the Jupiter Beach Resort & Spa. The beach curves in a jagged arc to the southeast, in front of a state park that begins where the resort's property ends and extends for about three miles. To the north are three condominiums, then more state parkland ending in the Jupiter Inlet, about three-quarters of a mile away. In a feature unusual for Florida, there are rock formations protruding from the sand at the water's edge, for about an eighth of a mile to the south and a quarter of a mile to the north of the hotel.

It is cool today, for this time of year, in this place. But the water is warm enough for pleasant wading in the surf. The waves are rolling in from the northeast, carrying wet-suited surfers obliquely to the shore. The sea has been rough during the three days that we have been here. As the tide rises, the surf's assaults sculpt the coquina, sending foaming spray ricocheting skyward. Today, gray is missing from the seashore's palette. The robin's egg blue and white of the sky and the browns of the beach frame the greens, blues, and whites of the ocean.

Since I was a child, the ever-changing seashore has kept its primeval hold on me—through the sights, sounds, and smells of the ocean pounding against the shore; through the feel of the sand beneath my feet, now soft, now firm, now wet, now dry, now warm, now cool, now coarse, now smooth; through the air's transmission to my skin of the sun's warmth, to my eyes of the sun's glare, to my body of the wind's pressure; through the incessant waves, each a life cycle, no two alike, remaking the beach with each advance and retreat, erasing the brief imprint of my footsteps; and through the symbiotic life-forms in the sea, in the air, and on the shore. The sea itself remains familiar and mysterious to me. Even when calm, danger lurks in its depths, and potential for crushing violence remains stored in its waves. By day, I am mesmerized by the seashore's stupendous juxtaposition of earth, sky, and water. At night, I am comforted by the surf's roar and pounding, the planet's slow, irregular pulse.

Some say that life on Earth was born in, and emerged from, the sea. Certainly, life cycles seem as natural to me at the seashore as they did in East Africa or in the Galapagos. I take some solace in thinking that my ashes will be cast into the ocean's vastness, whence some may reemerge on the shore.

THE RAINBOW'S ARCH

On Sunday, February 14, in a simple ceremony, my family interred my father's ashes in the little garden on the grounds of St. Mark's Episcopal Church in Jacksonville. Dad and Mother donated the gate to the garden, as well as one of the church's windows. Dad attended St. Mark's for decades, until he and Mother moved to Vicar's Landing, a retirement community in Ponte Vedra. The land on which Dad's father's house stood, and in which Dad had lived for a short time, is now part of St. Mark's campus. Dad also did some of the architectural work at St. Mark's, including designing the rectory.

On the way to St. Mark's, Susan and I drove past the turnoff to Lakeshore Junior High School, which my brother and sister and I attended, and which Dad designed. After the ceremony, we drove two blocks from St. Mark's to pass by Ortega Elementary School, where Dad designed the classroom additions in which my brother and sister attended their first- and second-grade classes, and—having skipped first grade (a bad decision, by the way)—I attended second-grade classes. We then drove by the 1939 vintage home on Ortega Boulevard in which my brother and sister and I were raised, and which Dad designed, to a lunch hosted by Mother at Timuquana Country Club, which, as I have noted previously, Dad also designed.

Yesterday, in Ponte Vedra, we held the church service in memory of Dad, and the reception afterwards, at Christ Episcopal Church. Dad attended that church for the many summers that he spent at Ponte Vedra, where, in the mid-1930s, he designed most of the original houses and, over the ensuing six decades, many others, including the two that he designed for us. Christ Episcopal Church itself was designed by Wellington Cummer II, a friend and colleague of Dad's, and the father of my late roommate, Chris, at the Hill School. When Mr. Cummer left the practice of architecture to enter his family's business, Dad drafted the architectural plans for Christ Episcopal Church based on Mr. Cummer's design.

Because Dad outlived almost all of his contemporaries, it was gratifying that the memorial was well attended. While the service was traditional and poignant, it was beautifully executed and was indeed a celebration of Dad's life. As my brother pointed out in his remembrance, Dad always greeted each day with optimism. In late March or early April 1513, Don Juan Ponce de León may have landed on the shore that we now call Ponte Vedra Beach, in search of the Fountain of Youth. Although Ponce de León never found it, I believe that Dad did.

Afterwards, at the reception, Mother insisted on standing, for as long as she could, to thank Dad's and her friends for their presence. After the reception, as I walked outside the church into a gentle rain, one of the clergy hurried over to me and pointed upwards: A rainbow's arch spanned the sky.

ISLAND HOPPING

All things considered there only two kinds of men in the world—
those that stay at home and those that do not.

—Rudyard Kipling

A MONTH DOWN UNDER

On Monday, February 22, Susan and I are heading Down Under. We are scheduled to arrive back in New York on March 22. Although we have long wanted a sojourn in Australia (or Oz, as we are learning to refer to it) and New Zealand, we thought that Australasia was too far away and too vast to visit while I was still working. Now we think: if not now, when?

I confess to some anxiety about taking this trip. Perhaps because of my sudden decline a year ago while we were in South America? Because of the increased risk of developing blood clots on long flights? (I've had two blood clots since I was diagnosed with cancer, and stroke is one of my greatest fears.) Because of the logistics of getting the airlines to permit me to take into their main cabins the syringes of Fragmin, a fragile blood-thinner that cannot be exposed to the temperatures of their luggage compartments? Perhaps out of concern that we may have scheduled an itinerary that will outpace my stamina? Because, by taking this trip, I am almost certainly foreclosing the opportunity to go to other exotic places that are in season and that we also postponed visiting while I was working, such as Egypt, South Africa, Patagonia, and Costa Rica? But any trepidation is overridden by my awareness of how lucky I am to have this unexpected opportunity for extended travel.

As the trip nears, my anticipation grows. We plan to swim, snorkel, and go boating and deep-sea fishing in the waters of the Great Barrier Reef; walk in parks, gardens, cities, and villages; visit museums and art galleries (I like Aboriginal art); go to the opera in Sydney; go racing in Flemington on the second most important day of racing in Australia each year; drive the Great Ocean Road from Melbourne; visit zoos and otherwise view exotic flora and fauna; visit stud farms and wineries; play golf (my balky right shoulder permitting) at the Royal Sydney Golf Club, as guests of a friend; climb a glacier; go boating in fiords; fly in helicopters and small planes over spectacular scenery; stay in good hotels and resorts; eat well; and drink the local wine and beer. Of course, Oz alone is so vast that we will not get to do many things that we would much like to do, including visit the bush and the outback, as well as Tasmania and coastal cities that are not on our itinerary.

For the long flight Down Under, I am taking along two books that I read about a quarter of a century ago: Robert Hughes's epic of Australia's founding, *The Fatal Shore*, and Bruce Chatwin's account of Aboriginals' Dreamtime creation myths, *The Songlines*. I haven't figured out yet what to read on the flight back to New York. Maybe I'll just try to catch up on the news.

FEBRUARY 26, 2010
FRIDAY 6:43 PM

ISLAND HOPPING

Auckland, New Zealand, is eighteen hours ahead of New York time. We spent much of the day on Waiheke Island, as Sally Peters and Hal Davidson, friends who had recently been there, had recommended. We took the public ferry from the center of Auckland, a thirty-five–minute trip. Thanks to rain squalls and vigorous winds, we were the only passengers avid enough to brave the open roof-deck of the ferry.

The weather cleared magically and conveniently as we reached Waiheke Island. We walked up a gradually inclining road for twenty-five minutes to reach our luncheon destination, the Cable Bay Vineyards restaurant, which is located on the eastern edge of the site of a Maori fortification, on a commanding hilltop now decorated with modern sculptures. Both the lightly oaked Cable

Bay chardonnay and the food were top quality, judging from the bread, greens, crème brûlée, and local oysters. The view west through the elegant, modern restaurant's open doors and expanse of plate glass is unforgettable—through clear air, over hilly harvest-time vineyards, brown and green pastures, and open sea, to the skyline of Auckland, under bright blue skies decorated with wisps of white clouds. After lunch, we toured the rest of the island by taxi. Waiheke's coastline is a series of protected coves, with pockets of shark- and coral-free, pleasant beaches, with white sands on the beaches on the north side of the island. As measured in fractals, Waiheke's coastline is quite long, not only in relation to its landmass. Waiheke is now being gentrified. There are newly planted vineyards (some of the most steeply pitched that I have ever seen), and a few fine residential properties. Increasingly, commuters travel to jobs in Auckland. The island was a hippie colony forty years ago. Throughout the world, hippies have been pioneers in finding great places to live.

This morning, at about 4:30 a.m., we left Mollies, our delightful hotel in Auckland, to fly five and a half hours to Cairns, Australia. Now in Cairns, I am writing this entry from the lounge of Hinterland Aviation, which is the air charter service that will fly us, in a twin-propeller Cessna 402, about a hundred miles north to Lizard Island, in the midst of the Great Barrier Reef. It is monsoon season, and it is currently raining hard. We are told that our flight will not be delayed by the weather and that conditions at the moment are perfect on Lizard Island. As it is off-season and Americans are not traveling to Australia as much as they did before the recession and the decline of the U.S. dollar, there will be only about a half dozen other couples on the island when we arrive—versus a capacity of about eighty guests. The resort shares the island with a National Park research station. The island is out of range of satellite service for cell phones. Inaccessibility is one of its charms.

THE LIZARD

"Are you off to the Lizard?"

The pilot introduced himself in the lounge and gave us safety instructions. We took off with heavy rain pounding the windscreen in front of the pilot and no visibility out of the windows to our sides. Twenty minutes later, we emerged into clear skies. The lapis lazuli of the sea below was punctuated by the mottled green, brown, and blue turquoises of the Great Barrier Reef.

An hour after departure, the pilot banked the plane around the largest, tallest island that we had seen on the flight, then landed on an incongruous asphalt runway running through the tropical landscape. For the next four days, the weather remained ideal for our purposes: no rain and smooth seas.

We snorkeled, boated in small skiffs, and hiked as our mood directed. One afternoon, we were taken to a stretch of the outer reef for snorkeling. On another, we trolled with lures for fish,

landing three. One of the catches turned out to be as much of a battle as I could handle when the butt of the rod snapped, after which I had to push down and pull back simultaneously with my left forearm to stabilize the rod while maintaining pressure on the fish, a giant trevally. The next morning, I found bruises along my left forearm from internal hemorrhaging of muscle. (I bleed more easily now that I have to inject myself each day with the blood-thinner, Fragmin.)

Our room was on the highest point of the resort. Its open balcony's view of the resort's anchorage had just about every element other than snow-capped mountains that one could put on a wish list: elevation, a view through tropical foliage swaying in the breeze, a small island, a few boats at anchor, a crescent of white beach, and a rocky promontory at the far end of the beach. Lying in a hammock, I was so tranquilized by this view and its associated sounds—rustling leaves, splashing surf, bird calls and echoes of bird calls—that ambition and restlessness left me. We experienced one of the wonders of the world, the Great Barrier Reef, under idyllic circumstances. Lizard Island was a very good place to celebrate Susan's birthday, on one of four of the best days of our lives.

MELBOURNE

Tonight is our fifth and last night in Melbourne. We are sorry to be leaving this attractive, well-organized, prosperous, sophisticated, and congenial city that prides itself on its excellent food and wine. We hired a driver to take us on the Great Ocean Road southwest of Melbourne until it turns inland from the coast—at which point, we flew in a small, single-engine plane along the shore, past the famous rock formations known as the *Twelve Apostles* (now eroded to about Nine-and-a-Half-Apostles).

This morning the sun was shining, and the temperature moderate: perfect weather for "Super Saturday," the big day of racing at Flemington, the impressive racecourse on the outskirts of Melbourne. Suddenly, following the sixth of the nine races carded for the day, the skies darkened, winds tumbled lawn furniture, and a diagonal wall of water cut off vision. Hailstones the size of baseballs bounced off the ground, then coating it white while emitting a low-lying fog. Several horses broke free of their handlers and injured themselves. The lights in the racetrack flickered on and off, and the tote board in the infield went blank. The stewards canceled the remaining races, but the crowd could not depart. The surrounding roads were flooded; the roof of a rail station had collapsed, and rail service was suspended.

The patrons took their hardship calmly, continuing to eat and drink and chat with one another while betting on races at other tracks. A local bloodstock agent, Vin Cox, had organized a table for Susan and me in the track's Terrace Room, a restaurant that turned out to be formal by American standards. I had worn the best clothes that I had with me—a double-breasted blue blazer, charcoal-gray slacks, white shirt, conservative tie—and felt underdressed. With no way to go anywhere, we settled in for high tea, and I found a Group One race in Sydney on which to bet. My selection, Theseo, won by a nose, at 8.5-to-1 odds, which put me ahead on the day, just as our driver called to say that he was waiting for us at the entrance with an umbrella.

When abroad, it is difficult to know whether a phenomenon is out of the ordinary while it is unfolding. During the storm, I thought that the weather in Melbourne must be a bit volatile. According to the newspapers, the storm was the most violent in Melbourne in nearly forty years.

MARCH 14, 2010
SUNDAY 4:56 PM

SYDNEY AND THE HUNTER

The weather was clear and warm during our stay in Sydney. Brief holiday stays are always vulnerable to weather. But we were especially dependent on conditions for March 9, as we had booked a helicopter for the day to take us, along with Vin Cox's assistant, Garry Cuddy, north to Hunter Valley ("The Hunter") to visit the four most important stud farms in Australia—Widden, Darley, Coolmore, and Arrowfield. March 9 was a picture-perfect day for flying. We loved our hour-long flights to and from The Hunter over the wilderness of the rugged, dark green, national park. Thanks to Vin Cox and Garry Cuddy, we were well received at these world-class farms, each of which operates on a staggering scale. Australian thoroughbreds are sounder looking than their American and European counterparts, with bigger bone and more solid feet. The box stall of the leading sire in Australia, Redoute's

Choice, is the size of a small, indoor riding ring. A highlight of the day for Susan was pulling Redoute's Choice's tongue.

In Sydney, our room at the Park Hyatt had a picture-postcard view overlooking the Opera House and the harbor. Our stay was greatly enhanced by a friend from New York, Lee Edwards, a retired professor of art history who lived the first nineteen years of her life in Sydney and now spends her Northern Hemisphere winters there. As I have long been interested in Aboriginal art, seeing it in museums and galleries with Lee was a prized opportunity. She had great seats at the opera, where we saw a fine production of *La Traviata* in the surprisingly intimate, fifteen hundred–seat auditorium. She took us not only to the public parks, beaches, and restaurants for which Sydney is noted, but also to her golf club, the Royal Sydney, and to her yacht club, the Royal Prince Edward, which was founded by her grandfather. On our own, Susan and I went shopping and took a ferry to visit the zoo.

On March 12, we flew from Sydney to Christchurch, New Zealand. To summarize what I think about Australia in general, and Sydney in particular: If I were young, just starting out, with my life in front of me, I would move to Sydney.

MARCH 17, 2010
WEDNESDAY 6:36 PM

ROSES AND BUNGEE JUMPING

Because our flight to Christchurch arrived after midnight, we spent a quiet first day in New Zealand's so-called Garden City. We walked to the small city's permanent sights rather than visit the Ellerslie Flower Show, which was being held for the first time in Christchurch. In Hagley Park, purportedly the third largest city park in the world, we meandered among magnificent trees and in the rose gardens. On our second and final day in Christchurch, we hired a driver to take us to a whaling village, Akaroa, founded by French settlers in 1840. After lunch at a French restaurant, we took

a small tour boat through the towering canyons of the Sound to its Pacific Ocean headlands, accompanied part of the way by rare Hector's dolphins, the smallest dolphins in the world.

The range of activities in Queenstown, where we spent four nights, is more spirited than in Christchurch and attracts a younger crowd for its bungee jumping, kayaking, jet boating, jet skiing, whitewater rafting, extreme rafting, whitewater sledging, river boarding, river surfing, fixed-wing flying, helicopter flying, skiing, heliskiing, helibiking, sky swinging, canyon swinging, mountain biking, vertigo mountain biking, gliding, paragliding, tandem paragliding, hang gliding, tandem hang gliding, stunt flying, ballooning, skydiving and tandem skydiving, gondola riding, luge, indoor ice skating, lawn bowling, clay-target shooting, horse trekking, back-country saddle expeditions, fishing, tennis, golf, frisbee golf, miniature golf, sheep shearing, visits to wineries, and pub crawling. Bungee jumping started here; the original site remains in operation and observes each anniversary of the first jump by offering free jumps to nudes. Most of those free jumpers are said to be young women.

I talked Susan out of the tandem skydiving from fifteen thousand feet. We contented ourselves with such day trips as visiting wineries in Central Otago, noted for their Pinot Noirs; and by taking a helicopter, on a perfect day without a cloud in the sky, over the mountains and glaciers of the Southern Alps to Milford Sound, where we took a boat trip through the fjord. Our run of good luck with the weather finally broke, and a trip by jet boat into deep forest rivers (including sites featured in *The Lord of the Rings* movies) was canceled. We hiked the outskirts of Queenstown instead, including stumbling on an exquisite rose garden at the top of a public park, which was on a peninsula in Lake Wakatipu. The sunlight is unfiltered by pollution. Pure light illuminates the terrain's glaciers, mountains, hills, valleys, forests, rain forests, and vineyards and makes the pristine freshwater rivers and lakes and the salty sounds, fjords, and sea sparkle. When it is cloudy, the slopes change colors as the shadows play across their surfaces.

MARCH 19, 2010
FRIDAY 10:49 PM

THE PAINTER'S PALETTE

From Queenstown, we flew in a turboprop along the Southern Alps to Wellington, at the southern tip of the North Island of New Zealand. In Wellington, we changed to a small plane to fly to Taupo, in the center of the North Island, where we stayed in the Huka Lodge, justly famous.

For our main activity today, we hired a helicopter to pick us up at the Huka Lodge and take us sightseeing for three hours. Taupo is lush and green, but our destination was White Island, in the Pacific Ocean—an active volcano twenty-five miles offshore. White Island is a classical depiction of Hell. Its volcanic slopes focus the heat of the sun. The bowels of the Earth defecate black sludge and sulfuric fumes through bubbling cauldrons. A puce-green sulfuric

lake backs up against a slope. The Earth's crust sporadically bursts open wounds, spewing ash and stones (we were equipped with government-mandated hardhats and gas masks). The moonscape terrain is coated with fresh, bright, mineral paints—multicolored vomit heaved from the guts of the Earth.

On our way to White Island, we flew above a tectonic plate fault line that registers about a thousand earthquakes per year. The separating plates are marked by geothermal pools, the coloration of each depending on its mix of minerals. One series of juxtaposed pools is called The Painter's Palette. We landed above the clouds on an inactive volcano, Tarawera, that last erupted in 1886 with significant loss of life. The local Maori had long buried their chiefs on Tarawera and blamed the European settlers for causing the eruption by setting foot on the mountaintop and upsetting the gods. Nowadays, the helicopter pilots pay a fee to the Maori for landing on Tarawera, which seems to placate the gods.

On our return from White Island, we landed on a deserted beach and ate a sumptuous picnic lunch from the Huka Lodge. This afternoon and tomorrow, we will go on hikes. On Monday, March 22, we will have a driver take us to three of the leading stud farms in New Zealand—Waikato, Rich Hill, and Cambridge—on our way to Auckland, for a last dinner in New Zealand. We will fly to Los Angeles, then to New York, with a scheduled arrival at 6:00 a.m. on Tuesday the 23rd.

RELATIVE TIME

It is not death, it is dying that alarms me.

—Michel de Montaigne

RESUMING CHEMOTHERAPY

New York City has been unseasonably cold all day, beset with heavy rain and high winds—nasty weather. At MSKCC this morning, Dr. Saltz told me that my cancer is growing: "Everything is a bit larger" than it was a couple of months ago. The growth of the liver tumors is "just starting to interfere with bile output."

Dr. Saltz said that I have a "closing window" to try to arrest temporarily the growth and shrink the tumors. He recommended that I undergo the chemotherapy called FOLFOX, even though a not-infrequent side effect of the oxaliplatin in this cocktail is neuropathy. I am weighing in my mind that oxaliplatin is much more likely to cause neuropathy in those like me who have Raynaud's syndrome. Like many chemotherapies, FOLFOX has many side effects and potential side effects. For example, most people who are treated with FOLFOX experience a change in the way food tastes. Usually, but not always, this alteration in the sense of taste reverts to normal after halting the cycles of FOLFOX.

Dr. Saltz said that he hopes that, at some point, I will once again go without chemotherapy, for positive reasons. He emphasized that he will be continually reevaluating my treatment, and that oncology is the art of "Plan B." He said that he thinks that I will have periods in each two-week cycle of chemotherapy during which I can get exercise and otherwise enjoy life.

This afternoon, over a period of a little over two hours, I received my first dose of FOLFOX. The risk/reward ratio of beginning this round of FOLFOX seems rational to me. I don't think that I have fallen into the trap—so far, at least—of desperately flailing about, mindlessly permitting poisons to be pumped into me that offer no real hope of delaying the inevitable.

ELEMENTS OF STYLE

When I was growing up in Jacksonville, Florida, its public schools were in chaos. My school had double sessions during the tenth grade, my last year of schooling in Jacksonville. I was never taught the rules of grammar, and I don't know them to this day. My parents sent me to a prep school—the Hill School in Pottstown, Pennsylvania—for my last two years of high school, so that I would have a chance at getting into a good undergraduate school. (My brother had attended the Hill for three years, and he was matriculating at Princeton.) The Hill's headmaster was an English teacher, and the school put a lot of emphasis on writing. We had to write an essay for English class each week, and I tried hard to avoid making the same mistake twice. *The Elements of Style* by Strunk and White was my Hill School grammatical bible. I marvel at what I consider to be good writing, and I am bothered to the point of distraction by what I consider to be bad writing. Sloppy writing with unintended ambiguities frustrates me. After reading great literature in prep school and college, most of the reading assignments in graduate business school were an ordeal for me.

I once read something by, or attributed to, Peter Drucker, to the effect that different people learn in different ways and that he himself learned by writing, not by listening or reading. It dawned on me that maybe I learn best the same way. When I am writing, I concentrate more, and my mind engages more, than when I am reading or listening. But I have written little. No wonder that I have learned so little over the years.

PECAN PIE

A friend from South Carolina, Pat Martin, who is married to my great friend since childhood, George Martin, sent me one of her prize-winning pecan pies for Easter. My grandmother Kiker—my mother's mother—also made wonderful pecan pies. My grandparents had a big pecan tree in their front yard in Reidsville, North Carolina. I loved to climb that tree, and, in great expectation of the pecan pie to follow, I would eagerly gather pecans from it, crack them, and pick their meat.

A MUSEUM OF REGRETS

Whenever I read an interview of some prominent man who says towards the end of his life that he has no regrets, that he wouldn't change a thing, I am puzzled. If he ever played golf, didn't he ever wish for a mulligan? Did he ever make investments? He certainly did not buy racehorses. I just don't get it. How does one mature and learn to be a better person without regrets? An expression that has always made sense to me is, "Life is a museum of regrets."

LIMBO

My first treatment with FOLFOX two weeks ago not only resulted in a potpourri of relatively mild, expected side effects, but also a couple of unexpected side effects, including a precipitous drop in my platelet count. This "very rare" reaction coincides with an enlargement of my spleen. My enlarged spleen may or may not be sequestering some or all of the platelets that are absent from my bloodstream. Because of my reduced platelet count, Dr. Saltz reduced by half the amount of Fragmin that I inject into myself each day as a measure against blood clots. I go daily to MSKCC to have my platelet level treated.

I had been scheduled today for my second treatment with FOLFOX, but my platelet count had not risen sufficiently to allow the treatment to proceed. The current plan is for me to return to MSKCC in a week to receive a reduced dosage of FOLFOX, if my platelet count has risen sufficiently. Meanwhile, my platelet count is now high enough both to end daily testing at MSKCC and to resume injecting myself with full doses of Fragmin. Instead of being treated for cancer, I am recovering from chemotherapy.

WHY NANOTECHNOLOGY?

A few months ago, a friend at M.I.T. gave me a copy of a great book on nanotechnology, *No Small Matter: Science on the Nanoscale* by Felice C. Frankel and George M. Whitesides. According to the authors, "We will not know the ultimate impact of nanoscience and nanotechnology for many years. It is transforming the way science and engineering work together, and rewiring the sociology of technology; it is brokering the movement of knowledge among communities that barely knew of one another's existence." Lastly, they note: "Also, it's already very, very cool, and very useful."

APRIL 15, 2010

THURSDAY 11:51 AM

RELATIVE TIME

A friend, who recently underwent surgery about a month after receiving his cancer diagnosis, remarked in an e-mail to me a couple of days before his surgery that it had been a long month. My experience with time has been similar. It has been less than fourteen months since my cancer diagnosis, but it seems to me that the diagnosis was rendered years ago—five years, ten years ago, maybe longer. I wonder if this perception that time is slowing is common for people living with cancer or terminal disease?

It has always been my impression that healthy people perceive that time is accelerating as they age. The explanation that I have heard for this perception of accelerating time is that a year is twenty percent of a five-year-old's life, but only two percent of a fifty-year-old's life. Although this explanation seems plausible, it seems rather pat, and I have never been satisfied by it. And why may the correct explanation, whatever it is, not apply to those living with cancer or terminal disease?

FUN IS GOOD

My friend, Matthew Stevenson, recounts in his latest book, *Remembering the Twentieth Century Limited*, that the St. Paul Saints, a minor-league baseball team that was run by the actor Bill Murray, had as its motto, "Fun Is Good." This delightful motto reminded me of something more solemn that a French friend—a woman—told Susan and me some three decades ago: "It is important to have fun, because life is not funny."

LEGACIES

A few days ago, Mersana Therapeutics, Inc., and Teva Pharmaceutical Industries Ltd. announced an agreement in which Teva would pay, if all milestones were met, at least $334 million to Mersana in exchange for rights to develop and market Mersana's XMT-1107, which combines Mersana's Fleximer with a drug derived from fumagillin. Fleximer is Mersana's proprietary, nanoscale polymer that can be linked to small molecules and biotech drugs across therapeutic categories to enhance their delivery—with the goal of improving effectiveness and safety.

Until I retired from Harris & Harris Group, I was a director of Mersana. When I read the news about the Teva deal, I e-mailed Mersana's Chief Executive, Julie Olson, and Chief Operating Officer, Pete Leone, to congratulate them. Julie made me feel good by writing in her e-mail in reply, "We wouldn't be here had it not been for all of your support and belief in the technology and the team."

NOT GIVING UP

This morning, my platelet count was high enough for Dr. Saltz to recommend that I resume chemotherapy with a reduced dosage of FOLFOX. Summing up my remaining therapeutic options, he said that we don't have many cards left to play. At my request, Dr, Saltz outlined what I gathered to be the two remaining potential alternatives. One alternative is a chemotherapy that I could take only if the results—which he is awaiting—of a genetic test indicate that it is suitable for me. This chemotherapy, which couples irinotecan with Erbitux, usually has side effects that sound hideous to me; when it works, it always has these side effects. I will not make a decision about this chemotherapy unless it becomes an actual alternative. At this moment, it seems to me that subjecting myself to this chemotherapy would be inconsistent with my desire for quality of life, as opposed to quantity of life.

The second alternative is FOLFIRI, the chemotherapy that I underwent last year. FOLFIRI could not be expected to have much effect on my cancer now; my current cancer cells have undergone a Darwinian selection process for resistance to this chemotherapy.

This morning, I learned also from Dr. Saltz that I have developed a surgical hernia—a condition unrelated to an inguinal hernia. Like any other hernia, this surgical hernia could only be corrected through more surgery. Even if a surgeon were willing to perform, and if I were willing to undergo, such elective surgery, I could not do so without halting chemotherapy. Contrary to all instinct, I must passively accept permanent, uncomfortable disfigurement from this surgical hernia as part of the collateral damage of my cancer treatment.

Dr. Saltz cautioned Susan and me that my disease could take a sudden turn for the worse. In response to a question from Susan, he indicated that I should have my affairs in order. In response to a question from me, he said that although I could see the hospice consultation service at MSKCC at any time, he did not think that it was necessary at this time. I was startled and gained new perspective when he added that I could consider my chemotherapy as palliative care.

Sic incurable cancer.

ON THE INSIDE

The spirited horse, which will try to win
the race of its own accord,
will run even faster if encouraged.

—Ovid

A GOOD TRADE

This morning, Todd Pletcher, the trainer of Eskendereya, the overwhelming favorite for this Saturday's Kentucky Derby, announced that the horse injured a foreleg and will not run. Similarly, last year's Derby favorite, I Want Revenge, was scratched on the eve of the race because of injury. To me, Eskendereya appeared not only likely to be the shortest-priced favorite in recent years to contest the Kentucky Derby, but also likely to be the first horse to win the Triple Crown since Affirmed in 1978.

Like Grand Prix race cars, thoroughbred horses have always been fragile. In order to be competitive, both have to achieve an exquisite balance between power and weight and between speed and sufficient structural soundness to be fast enough to win and still complete the race.

In the United States, thoroughbred horses are becoming even more fragile, making fewer and fewer annual and career starts. The gross earning power of a stable of racehorses is partially a function of the number of races that they contest. All else being equal, a horse owner has to maintain more and more horses to generate the same gross revenues. An owner has to have great passion for the game to stay in it. Few stay for long, although since 1973, Susan and I have owned   on our own and with partners—a number of horses.

Last year, we sold one horse and found a home for another that was no longer racing sound. Because of my diagnosis, I did not go to the yearling sales last summer to replenish our stock. We now have only three horses: Hot Money, Mustang Island, and Backslider.

It was tempting to ask Christophe to race at least one of our remaining horses—Mustang Island, a sound horse coming off a winning race—in Florida this winter, so that I could have the pleasure of watching him train and race. But I knew that, if Mustang Island ran in the winter, he would not be at his peak for a spring and summer campaign on the grass courses of Belmont Park and Saratoga. I never broached with Christophe the subject of running Mustang Island in the winter. There is a right way and a wrong way to do most things in racing, and I would find no satisfaction in doing things the wrong way.

Like Mustang Island, Hot Money is a four-year-old gelding. He will be entered at Belmont in a six-furlong allowance race for "New York–breds," to be run on the turf on May 5. Christophe thinks that Hot Money is at his best on firm turf, so we will now start worrying about rain. According to Christophe, Mustang Island is not in peak condition, but could possibly be ready for a race on May 8.

Although Backslider is a three-year-old, he is unraced. He has not yet—to use a horseman's expression—come to hand, but he might be ready to run by about the end of May.

If I had never owned and bred racehorses, my net worth would be higher. But I would have had less pleasure in life. Most important, racing has been a family activity. I made a good trade.

APRIL 28, 2010
WEDNESDAY 11:28 AM

COLLATERAL DAMAGE

Yesterday's Senate hearings—in which lawyered-up executives of Goldman Sachs were grilled on television—were great theater. The hearings even produced comic moments, such as a senator from Nevada taking umbrage at comparisons between Las Vegas's casinos and Goldman Sachs.

Whatever new financial regulation emerges from Congress will not be so funny, with all sorts of unpredictable, unintended consequences. To me at least three of these unintended consequences are predictable. Government bureaucracy and costs of financing in the United States will increase; the new legislation will widen the relative advantage of the largest firms, like Goldman Sachs, as regulation favors the incumbents; and financial practices that are inhibited here will move abroad.

The moribund venture capital industry is an example of unintended consequences of government regulation spawned during political frenzies. A decade ago, the United States's vibrant venture capital industry was the world's envy. Since 1980, companies no more than five years old have accounted for approximately all of net new jobs. On balance, older companies have hired no one. A decade ago, in technology-driven areas like Silicon Valley, Boston, and Austin, venture capital financed about a third of all start-up companies. Professional venture capitalists backed many of what are now the most important companies in the United States, including Apple.

All but a handful of venture capital general partnerships are struggling to replace the capital in their expiring funds. Over the last decade, the venture capital industry's returns to its limited partners have been abysmal, and limited partners have been responding by cutting back or eliminating their investments in venture capital. The venture capital industry is losing the capacity to back start-ups, notably in capital-intensive industries.

The government's inadvertent destruction of the infrastructure that supported venture capital began in the late 1990s with decimalization of stock trading. The government also took away the market-maker's information advantage by requiring market-makers to display their best quotes publicly, rather than exclusively to other market-makers. As a result of these reforms, market-makers could no longer make a profit dealing in any but the largest companies' securities, and they ceased making markets in the less active securities of smaller capitalization companies.

Next, in the name of reform, the government mindlessly completed the destruction of research coverage of smaller capitalization companies by banning participation of securities analysts in investment banking revenues. Analysts were also restricted in helping to market initial public offerings. If an institution bought an initial public offering, it would have to do so with limited analytic assistance in understanding the new company, with little or no subsequent research coverage of the company, and with no market-maker standing ready to commit capital to facilitate any subsequent buying or selling of the new company's shares.

Just in case a smaller company somehow manages to overcome these obstacles and go public, it now has to deal with the requirements of the Sarbanes-Oxley Act, a poorly drafted piece of legislation passed in the dead of night during the political frenzy over WorldCom and Enron. Congress and the Securities and Exchange Commission had no idea what it would cost companies to comply with Sarbanes-Oxley. The deleterious result for smaller companies was that it became prohibitively expensive for most of them to function as publicly traded companies. An unintended consequence of these regulations is to entrench large companies by building a barrier blocking smaller companies from access to the public capital markets.

Ironically, the government historically extolled and passed legislation to help the venture capital industry and never targeted it for reform. From time to time, there will still be a few initial public offerings by small companies. More frequently, venture capital firms will sell their portfolio companies to larger companies. More rarely, a venture capital–backed company like Google will become so dynamic that it will stage a wildly successful initial public offering. But initial public offerings and the venture capital industry as we knew them are moribund. They were just collateral damage.

"EVEN GOD CAN'T HIT A ONE IRON"

My mother's father, Daddy Bill, died nearly a half century ago. One of our family stories about him concerned the instance in the mid-1930s when he loaned Bob Jones (popularly known, to his displeasure, as "Bobby" Jones) his golf clubs when Jones found himself at Augusta without his own clubs.

My father inherited that set of clubs from Daddy Bill and added its one iron to his own set of clubs. The rest of the clubs in Daddy Bill's set, rendered sacred relics in our minds by Bob Jones, were eventually lost or stolen. When my father stopped playing golf, he passed the one iron along to me. I found it nearly impossible to hit and donated it to the museum at Redtail Golf Club in Canada, at which I had playing privileges. There is no disgrace in not being able to hit a one iron. Lee Trevino, who had been struck by lightning on a golf course, said, while exiting a course after a lightning alert and waving a one iron over his head, "Even God can't hit a one iron."

My sister recently sent me some old newspaper clippings. One of them chronicles Daddy Bill's interaction with Bob Jones. It seems that the setting was not Augusta National, which had just been completed, but rather Forest Hills, in North Carolina. Jones was using borrowed clubs and was struggling with the driver in the set. At the ninth tee, Daddy Bill joined the small gallery following Jones's foursome and loaned Jones his driver. Jones hit Daddy Bill's driver so well that he even used it for his second shot on a par five.

How did the tale of Jones borrowing Daddy Bill's driver become so embellished in my family? Was it just the process of retelling over a period of years? My father was the keeper of golf stories in our family, and he loved to tell a good story. I don't know if my father was familiar with Mark Twain's dictum, "Never let the truth get in the way of a good story, unless you can't think of something better." Once my father told a version of a story that he liked, he stayed with that version until it became his memory of the event. If my father had been shown the newspaper article, I'm sure that he would have laughed and added to his repertoire my magnificent gift of the one iron, never touched by Jones, to the golf museum.

I played Augusta National once, as a guest of Ted Danforth. When we reached number twelve, the famous three par on the "Amen Corner," I didn't realize that the pin was in its treacherous "Sunday-at-the-Masters" position—the sucker pin placement. Ben Hogan said that if he ever birdied number twelve on Sunday at the Masters, you would know that he had missed his tee shot. Not knowing any better, I took dead aim at the pin, and my ball came to rest about two feet from the hole. Not only the other members of my foursome, but also the caddies, couldn't contain their laughter at my dumb luck. Striding across the Hogan Bridge spanning Rae's Creek on my way to the hole, all I could think about was the glory of making a birdie at number twelve at Augusta, with the pin in that storied location.

I missed the putt.

MAY 3, 2010
MONDAY 11:38 AM

ON THE INSIDE

Last Saturday's Kentucky Derby was won by Super Saver, ridden by Calvin Borel. Jockey Borel's winning ride was his third Kentucky Derby victory in the last four years—an unprecedented feat. In one of the winning years, Borel rode an extreme long shot. Each of Borel's Derby-winning rides has been a variation on a theme: Drop the horse to the inside rail at the beginning of the race, relax him, save ground throughout the race, then surge along the rail to the finish line, as the tiring horses in front drift out towards the center of the track.

Everyone knows what Borel is going to try to do. Nonetheless, in the Kentucky Derby, the other jockeys, who usually make a point of not being shown up by letting a horse pass them on the rail, let Borel through unimpeded. My only explanation is that the Kentucky Derby must be so important to the other jockeys that they are more concerned about getting the best out of their mounts than they are about reaffirming their territorial "rights."

Fortunately, I had a bet on Super Saver, who went off at 8-to-1 odds. That payoff almost got me back to even on my day's wagering.

MAY 4, 2010

TUESDAY 9:40 AM

MORTALITY AND IMMORTALITY

I have thought a lot about dying, and I have feared it greatly. I have not, however, dwelled on thoughts of being dead, except for its effect on my family. Since I was a small child, I have equated being dead with nothingness.

With respect to physical and spiritual mortality and immortality, we humans differ from other life forms in our acute knowledge of our mortality. This great awareness of the life cycle does nothing to free us from it. We are conceived, we are born, we live for a time, during which we may reproduce and nurture offspring, and we die. I do not believe that, other than through genetic contribution to progeny, there is literal life after death for us or for other forms of life.

Human beings have a unique power, through our big brains and technology, to affect planet Earth and all forms of life that it supports, long after our individual lives cease. Unfortunately, it is easier for a human being to wreak long-lasting damage than to create long-lasting good.

Eventually, both the human species and our planet will cease to exist. Thus, immortality seems to me to be a meaningful concept only if defined oxymoronically as having temporal limits coterminous with the survival of planet Earth or with the human species.

Our ideas may achieve immortality—if not ascribable, then through the legacy of their own DNA. Ideas interact with other ideas that give birth to other ideas, and so on. Hunting, gathering, toolmaking, know-how, language, oral tradition, music, farming, literature, religion, superstition, war, politics, science, teaching, law, technology, medicine, art, architecture, trade, business, finance, design, games, and all other human endeavors are based on, develop, and amplify ideas. In any human idea, invention, innovation, construction, writing, song, conversation, or other communication, there is potential immortality.

I do not believe that "things happen for a reason," or that a personal God has a plan for each of us, or indeed for the human species. The meaning that I found in my cancer was that my ordeal spurred me to try to open up to others, to learn more and grow faster during my remaining lifetime. And the kindness of others who learned of my cancer showed me the best side of humanity. Although I allow for the possibility of free will, I think that randomness determines our individual fates. It is good to be lucky, as I have been in so many ways in my life.

I am sixty-seven years old. In another sixty-seven years, there will be, at most, a few people who will say that they knew or remember me. In five hundred years, like all but a handful of my 6.8 billion living brethren, I will be traceable, if at all, only in genetic material or genealogical records. Nevertheless, my fleeting life, like that of any of my fellow humans, may leave in its wake contrails of immortality.

FROM CHEMOTHERAPY TO HORSE RACING

Yesterday, after reviewing my platelet levels and examining me, Dr. Saltz recommended that I undergo another infusion of FOLFOX. He said that he thought from his examination that the first two cycles of treatment with FOLFOX might be shrinking my tumors. I accepted his recommendation and commenced the third cycle of treatment.

If I feel well enough, I will go out to Belmont Park on Friday, to watch Hot Money run in the sixth race, and on Saturday, to watch Mustang Island in the tenth race. Based on Hot Money's performances on the turf, and the apparent quality of his opposition, he will probably be favored to win. Although he has been away from the races for a long time, Christophe is renowned amongst bettors for having his horses ready to run after long layoffs.

Mustang Island's prospects on Saturday are more difficult to assess. He is behind Hot Money in coming to hand this spring.

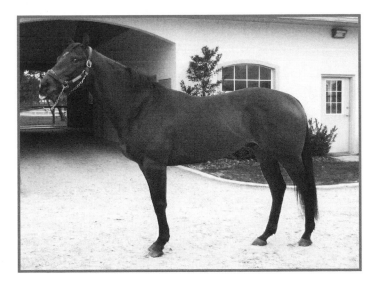

Unlike Hot Money, he is still shedding his winter coat, and he has had one or two fewer works than Hot Money. The talent of the competition in Mustang Island's race is more difficult to assess, and Mustang Island has never before been asked to sprint on the turf. As was Hot Money, Mustang Island will be ridden by Rajiv Maragh, a talented young jockey.

I think that the probable favorite in Mustang Island's race will be Onzain, which belongs to a friend of mine, Bill Punk. As owners and trainers say about their friends' horses, if I can't win it, I hope Bill does.

MAY 7, 2010
FRIDAY 6:11 PM

A DIFFERENT KIND OF PAIN

For the last few days, I have been not only beset by fatigue and other common reactions to chemotherapy, but also by relatively minor bone pain from Neulasta, a drug with which I am injected two days after undergoing chemotherapy to counteract chemotherapy's suppression of my production of white blood cell production.

Nevertheless, my anticipation of Hot Money's return to the races has been growing. I summoned my energy and went with Susan to Belmont Park to see him in the paddock and watch his race.

It was sunny and cool, with low humidity—perfect weather for racing, or just about anything else. The turf course was firm. Hot Money was calm in the paddock, fit without being lean, with a shiny summer coat—just the way one would like to see one's horse at the beginning of a racing season. In a field of eight, he was the 8-to-5 betting favorite.

After breaking alertly from the outside post position, Hot Money settled on the outside of the field, several lengths from the lead, as Maragh assessed their unfolding position in the race.

Nothing was opening up towards their inside. In order to avoid losing too much ground on the sweeping turn into the stretch, Maragh either had to pull Hot Money towards the back of the field or gun him towards the front. Maragh decided on the latter course of action, and Hot Money accelerated around the field, reaching the leader with a quarter of a mile to run.

At that moment I thought that Hot Money was going to draw away from the field and win in fine style. He had finished strongly in his morning works and his winning races. But his momentum faded, and he finished fourth. Although he seemed to be fine afterwards, it often takes a few days after a race for problems to manifest themselves.

Despite my reluctance to rationalize and make excuses, there are plausible explanations for Hot Money tiring that have nothing to do with his health. Perhaps he wasn't as fit as we thought after his long layoff? Jockeys are always convenient scapegoats for owners, trainers, and bettors, and one could be critical of Maragh's decision to unleash Hot Money so early in the race.

My hopes turn to Mustang Island's race tomorrow.

MAY 8, 2010
SATURDAY 8:02 PM

BETTING THE FAVORITE

Although I don't like to lose, I was content with Mustang Island's second-place finish in his race today as the 3-to-1 betting favorite in a field of twelve. The race was at the same distance and over the same surface—Belmont Park's inner turf course—as Hot Money's race yesterday. Because of a lack of rainfall, that course has been inordinately hard and fast. The last six sprint races over it, including Mustang Island's race today, have been won by the front-runner.

I had had a miserable night and day and did not make up my mind until mid-afternoon to go out to Belmont Park with Susan. During the hour or so that I spent with friends in the owners' boxes, visiting the paddock before the race, and watching the race, I felt no pain.

MAY 12, 2010

WEDNESDAY 11:51 AM

URGENCY IN LIMBO

The progression of my cancer drives my psychology. From the time of my diagnosis, I have accepted that I am incurable. Nevertheless, since being informed by Dr. Saltz that "we are running out of cards to play," time has sped up for me again. Living in limbo, I have a sense of urgency.

When I see a friend now, or watch on television an annual event like the Kentucky Derby, I am well aware that it may be for the last time. Although we renewed our subscription to our box at Saratoga—there is a years' long waiting list for those boxes—we have not rented a place to stay in Saratoga, which I never expect to see again.

Since my first CAT scan after my second surgery detected tumors in my liver, it has been clear that any additional treatments that I might receive would be palliative rather than curative. My stakes in the outcome of tests were lowered. From then on, whether a particular chemotherapy was efficacious or not would not be a matter of life or death.

As I look at alternatives, the final treatment that may make any sense for me to endure in order to try to extend my life is FOLFOX, the chemotherapy protocol that I am currently following. I am concerned about my ability to tolerate FOLFOX and about its efficacy for my cancer.

Dr. Saltz will not know whether FOLFOX is shrinking my tumors until my next CAT scan. By contrast, throughout each day and much of most nights, I am acutely aware of symptoms of my body's rebellions to the poisons in FOLFOX. .

Yesterday, in spite of my dosage of FOLFOX having been reduced by twenty percent, a blood test indicated that my platelet level has once again plummeted—a rare response to FOLFOX, according to the literature. If my platelet level does not rebound significantly by this coming Tuesday, May 18, I think that the best for which I can hope is that my next infusion of FOLFOX will merely be delayed.

As I can no longer predict how I will be feeling, I am no longer scheduling social engagements. Nonetheless, I agreed on short notice to have lunch with a friend who is leaving town for the summer. Since that pleasant lunch—at Raffaele, a neighborhood Italian restaurant—I have felt even better. Moreover, I am having a fairly productive day. Though it is raining and unseasonably cold, I am joyful.

MAY 15, 2010
SATURDAY 7:20 PM

MISSPENT YOUTH

The Preakness Stakes, the second leg of the Triple Crown, was run today in Baltimore at Pimlico Race Course. I have attended the Preakness four times, though not in recent years. Pimlico's dilapidated facility was known affectionately in its glory days, such as the era when Alfred Gywnne Vanderbilt was its president, as "Old Hilltop."

When I was a sophomore at Princeton, I won a daily double on a single bet at Pimlico on Preakness Day (i.e., I picked the winners of the first two races) at approximately 240-to-1 odds. The next Saturday I won at another track with another big daily double on a single bet. Thereafter, I had a small following of freshmen who would cut class to take my bets to the tracks. Ah, misspent youth.

I made a single bet on the winner of today's Preakness, and I was lucky enough to win the Exacta in the Preakness (i.e., I bet on the first two finishers in order) at 93-to-1 odds. Because I made two Exacta bets, not one, the odds on each of my Exacta bets were cut in half. As my bets on the Preakness were my only bets today, I did well—so well that I won about half of what I lost last week betting on my own horses, Hot Money and Mustang Island. Never take a tip from an owner on his own horse.

MAY 18, 2010

TUESDAY 4:29 PM

DRESS REHEARSAL

Today, I had a regularly scheduled appointment at MSKCC for blood tests and to be examined by Dr. Saltz and his associates. Then, if I were deemed to be in good enough shape to withstand another round of chemotherapy, I would go to MSKCC's chemotherapy suite. Although the blood tests indicated that my platelet count was at the borderline of being high enough for me to undergo another round of chemotherapy, Dr. Saltz decided not to proceed with the chemotherapy for at least a week.

Since early March, my belly has been growing more and more distended. It is now so large that the skin over it is stretched taut— like a woman's in her third trimester of pregnancy. Even though I have retained my sense of taste and a fairly good appetite, I can now eat only small meals because of the ensuing pain if I stretch my stomach more than minimally with food. I no longer exercise beyond taking walks of less than a mile, as the slight rotation of the torso that is part of each step generates both moderate internal pain and a shifting of the cloth of my shirt that irritates the hypersensitive skin stretched over my belly. The reduced caloric intake and lack of exercise are causing me to lose muscle mass steadily, yet I am gradually gaining weight. The only position in which I am now pain-free is sitting. My back hurts in other positions, presumably from supporting the growing weight of the belly.

Dr. Saltz and his colleagues say that they have not seen a similar case and have no idea why my belly is growing. They have scheduled me for more blood tests and a CAT scan and for appointments on Tuesday, May 25, with Dr. Jarnagin, the surgeon who operated on my liver, and with Dr. Saltz.

There is nothing that I can do about this mysterious complication other than be practical. For the last few weeks, I have been relegated to wearing low-cut jeans. Consequently, I decided to send my clothes to my tailor with instructions to enlarge the waistbands of the trousers and the waists of the jackets to the maximum extent that there is spare cloth in each garment. I hope that some of them turn out to be wearable. With no chemotherapy, I may feel good enough this week to go somewhere that requires clothing more formal than jeans.

MAY 19, 2010
WEDNESDAY 10:30 AM

THE GOOD FIGHT

The obituaries of people who die of cancer invariably state that the deceased died after a valiant "battle." While I have never doubted that anyone who died of cancer suffered and bore his or her ordeal heroically, I have found myself wondering if, in every case, the deceased fought his or her cancer, rather than sought some sort of accommodation with it.

If given an initial diagnosis of curable cancer, most cancer patients probably choose to battle their cancer for at least as long as it is still regarded as curable. Even though cancer treatments involve poisoning, cutting, and burning with both temporary and permanent side effects, some of which are disfiguring, and risks of potentially fatal complications, the possibility of cure is sufficiently motivating for patients to undergo almost any horrors and humiliations.

In my own experience, having been classified as incurable, then possibly curable, then again incurable, I think that the points of view of the curable and the incurable are different. For the incurable, all cancer treatments are palliative. Speaking as an incurable, I have no intention of doing anything voluntarily to drag out active dying. I share Frankl's view that "to suffer unnecessarily is masochistic rather than heroic."

For the incurable, the struggle may not be so much with cancer as with a medical establishment pushing in good faith, if sometimes mindlessly, the next treatment, as first one, then another prior treatment loses its efficacy. Fortunately, my oncologist is honoring my decision to seek quality rather than quantity of remaining life.

Since my diagnosis and re-diagnosis as incurable, my struggle has been primarily with myself, not with cancer or with the medical establishment. I must seek to use my remaining time wisely, to conduct myself appropriately, and to become a better person. If you read an obituary of me that says that I battled cancer bravely, you will know better.

BOOK IT

Yesterday, I signed a publishing agreement granting Cold Spring Harbor Laboratory Press (CSHLP) "rights to publish, and to license subsequent publication or other utilization by third parties, of a written manuscript entitled *Incurable: A Life After Diagnosis*." For the last few weeks, I have been working with an experienced team, consisting of an editor, Matthew Stevenson, a co-editor, Michael Martin, and a book designer, Nanette Stevenson (sister to Matthew), to convert the material from my blog into a book.

Although a number of friends have suggested to me since I started my blog that I should "write something," or at least rework the blog into a book, I knew that I would not have enough time to start a new book from scratch, and I couldn't be confident that I would have enough energy or time even to rework the blog into a book. Not until Matthew, an oft-published author and editor and a trusted friend of mine for decades, proposed to me a month or so ago that he and his team go to work with me on such a project did I stop dithering.

Matthew presented to me a solution that would ensure the production of a ready-to-print manuscript even if I could no longer work on it before it was finished. Matthew keeps reminding our team that "Every day counts," and we have made substantial progress. With the agreement with CSHLP signed and a friend, Al Perry, in place as a literary executor, I am now confident that the book will not only go forward, but also be limited in its quality only by my writing and thinking.

THE CLOSING WINDOW

Yesterday, as previously scheduled, I had more blood tests, and Susan and I met first with Dr. Jarnagin, the surgeon who operated both times on my liver, and then with Dr. Saltz. Dr. Jarnagin told us that last week's CAT scan showed that the FOLFOX was having "at best" no effect on the tumors. In response to a question of mine, he also said that, although there was no surgical intervention that would be efficacious against the tumor progression, drains could be inserted into my abdominal cavity to drain some of the fluid that keeps accumulating in my belly.

When we met with Dr. Saltz a few hours later, he confirmed that he was discontinuing my treatment with FOLFOX. He thinks that I now have tumors both in my abdominal cavity and in my stomach and that the fluid is cancer-related. Dr. Saltz indicated that he has only one other chemotherapeutic option for me to consider: Erbitux combined with irinotecan. I was previously exposed to irinotecan, as it is an ingredient in the FOLFIRI cocktail with which I was infused soon after my initial diagnosis. Although FOLFIRI set off a bewildering kaleidoscope of adverse reactions in my body, Dr. Saltz thought that I tolerated FOLFIRI relatively well. The more I think about my previous experience with irinotecan and study the literature on the irinotecan/Erbitux combination's probable and possible side effects, the more I marvel at patients' willingness to endure sufferings worthy of Job to try to stave off death.

If the irinotecan/Erbitux combination were to prove efficacious for me and its side effects were not life-threatening, I would receive it for about six months. If I recovered sufficiently from the side effects, I might have one last period in which I could go out into the world again.

On Sunday, I will undergo an invasive procedure under local anesthesia during which a drain will be inserted in my belly for some two hours. Dr. Saltz says that the chances of my getting another infection from the procedure—I got a total of four infections last year—are minimal and that the reduction in pressure should temporarily relieve some of the pain in my midsection. The fluid will rebuild to its current level, perhaps over forty-eight hours, perhaps over three or four weeks. As the abdominal pain is starting to awaken me at night (I intend to avoid taking pain medication for as long as possible), the potential benefit of the procedure seems to me to justify its purportedly minimal risk.

Until forced to make a decision sometime in the near future, I will continue to mull whether the hideous side effects of the irinotecan/Erbitux regimen would be compatible with my goal of maximizing the quality of my remaining days. Susan and I will meet tomorrow for the first time with MSKCC's palliative-care specialists.

MAY 27, 2010
THURSDAY 8:55 PM

ST. CHARLIE

My brother, who lives in Virginia, is visiting my mother. When I spoke over the telephone with her this evening, she told me that he had just read her a May 24th article, "A Good Trade," that was written about me by Ray Paulick, editor of the online thoroughbred horse publication, *The Paulick Report*. She exclaimed about Mr. Paulick's writing skills and about the dozen or so on-line responses to his article.

I told my mother that I certainly concurred with her about Ray Paulick's writing and that the responses to his article were more than kind about me. I added that I was, however, having a little trouble adjusting to my new role as *St. Charlie*. She started laughing and agreed that that would be a lot to live up to. She was still laughing as I bade her goodnight.

WHAT IS REQUIRED
OF ME NOW?

*It did not really matter what we expected from life,
but rather what life expected from us.*

—Viktor E. Frankl
MAN'S SEARCH FOR MEANING

WEIGHING THE ODDS

Yesterday, Susan and I met with Dr. Paul Glare, the Chief of MSKCC's Pain & Palliative Care Service, and with one of his colleagues, a nurse practitioner.

Dr. Glare volunteered to talk about my life expectancy. Previously, I have had to rely on the literature and conversations with non-MSKCC scientists to estimate how long I may have to live. According to Dr. Glare, if I wanted to enroll in hospice at this time, I would qualify. In other words, in the absence of more chemotherapy, it would be reasonable to assume that I would die within the next six months. Most of the colon-cancer patients sent to Dr. Glare—including patients who continue chemotherapy and those who do not—die in two to twelve months.

From Dr. Glare's specific examination of me and my records, his guess is that if I were to forgo additional chemotherapy, I would probably make it through this summer and die this fall. If I were to undergo additional chemotherapy and it were to prove efficacious, I would probably die later, but probably still within twelve months. He noted that third- and fourth-line therapies tend to be less efficacious than first- and second-line therapies. In my case, the first-line therapy, FOLFIRI, was moderately efficacious; the second-line therapy, FOLFOX, was not at all efficacious.

I am still thinking about the irinotecan/Erbitux chemotherapy and trying to decide if it would make sense for me, given my values and expected longevity. Dr. Saltz told me that when Erbitux is efficacious, patients always get a rash, but that patients for whom Erbitux is not efficacious may get the same rash. This rash looks like acne, may itch, and may extend from head to toe. While this rash is present, the skin has to be kept covered in ointments and cannot be exposed directly to sunlight. When Erbitux is discontinued, this rash usually, but not always, disappears, typically within a month.

Dr. Glare noticed that Susan was carrying a cloth bag from Coolmore Stud in Australia. He is Australian and said that he has kept his membership in the Australian Jockey Club. He asked if we have horses. I replied that we do and that we had a runner, Hot Money, entered at Belmont Park for the next day.

In Hot Money's race today, he was bet down to 5-to-1 odds. He ran well, closing fast at the end to gain third place. Before placing any bet, one should always weigh the odds.

MAY 29, 2010

SATURDAY 5:20 PM

WHAT IS REQUIRED OF ME NOW?

For many years, I have asked myself, "What is required of me now?" I do not know where I learned this saying, or if it originated in Buddhism or other religious and spiritual traditions. It may as well be a commonplace secular motivational tool.

Since I was diagnosed with cancer fifteen months ago, this question has continued to work. If I awaken in the morning in a comfortable position and dread the pain involved in getting out of bed, asking myself this question and then answering it (i.e., "I have to inject myself with Fragmin") prods me into action. When I cease to respond to this question, I will know that spirit is leaving me.

SUNDAY MORNING ON A LONG WEEKEND

Yesterday afternoon, I developed a fever for the first time since my surgeries. People undergoing chemotherapy frequently develop fevers, but I haven't had chemotherapy in several weeks. Because of the fever, when I reported to the Urgent Care Center to have fluid drained from my abdomen, the doctor on duty first checked for an active infection by testing my blood and urine and then ordered a chest X-ray. Finding no cause for the fever, she performed an ultrasonic examination to determine where to place the needle to drain as much of the fluid as possible.

Unfortunately, the fluid turned out to be behind the intestines and bowels, where it cannot be drained. Meanwhile, my midsection grows ever more distended and painful, both internally and, as the skin grows tighter, externally. Based on my weight gain, I estimate that about twenty pounds of fluid has accumulated so far. The doctor also reported that my liver is swelling and my colon thickening. She thought that it would be worthwhile for me to come back into the Urgent Care Center on Tuesday morning for her to perform another ultrasonic examination, just in case some of the fluid shifts to an accessible location. Afterward, I will attempt to meet with or speak with Drs. Saltz and Glare.

It is a sunny day in New York City, with low humidity. With many of Manhattan's residents away for the long holiday weekend, it is peaceful. Susan and I walked the dozen blocks from MSKCC's main campus to our home. The stretch along York Avenue in front of Rockefeller University is a particularly pleasant walk, paralleling an oasis of leafy trees harboring discreetly placed tennis courts.

SUSAN

Even though my life is being cut short, it has been my good fortune to have been married for over forty-two years to a loving wife. Susan is the sun around which our family revolves; not just our family—David, Elizabeth, and Elizabeth's daughter, Kayla— but her father, her father's friend, her brothers, sisters-in-law, nieces, and nephews.

Susan enabled me to focus on what I liked to do in my career and in our private business affairs by doing all of the thankless tasks for our whole family. I never saw a bill, much less paid it. I even signed our income-tax returns as the "innocent spouse." In our horse business, I picked out the young horses at the sales, studied pedi-grees, inspected and bought and sold horses, worked with the farm managers, veterinarians, and other professionals, and talked strategy with the trainers. She did the hard work, such as keeping the books, working out the depreciation schedules with the accountant, and purchasing the insurance coverages that I wanted. When we had a cat, Sparky, prone to biting and scratching, who developed lupus and diabetes, Susan was the one who put the prednisone pills in the back of his throat and injected him with insulin.

Engaging in all sorts of activities, from highbrow to lowbrow, from attending academic lectures and concerts of classical music and jazz concerts to playing and watching sports, we have had countless wonderful times as a couple and in the company of other couples, friends, and relatives. When we were first married, I had nothing but debts for my education, and we would attend Yankees' games on a budget of about ten dollars for the day—the centerfield bleacher seats were a dollar each. Knowing that our time together was growing short, we loved traveling last year to Florida and Paris and this year to New Zealand and Australia.

The most cherished memories of my life are times with our children when they were small. Susan is a superb cook and mother, and she made the holidays special—all that I had ever dreamed that holidays could be for a family. We had delightful vacations with the children. I remember celebrating David's fourth birthday at an outdoor table at a Michelin three-star restaurant in Provence, L'Oustau de Baumanière, where Elizabeth and David were more interested in the kitchen cats than in the haute cuisine. Each year, for the month of August, we rented a house in Saratoga Springs that left much to be desired, but it had the redeeming feature for young children of containing one of the town's few in-ground swimming pools.

My worries about being dead concern its effect on my family. Nothing can fill a departed family member's void, and I am sad for my loved ones. But I have the peace of mind of knowing that the sun of Susan's love will continue to provide our family with gravitational stability, warmth, and sunshine.

JUNE 1, 2010
TUESDAY 11:23 AM

TIME'S ARROW

When I was given the opportunity to choose elective surgeries and chemotherapies that would give me an estimated twenty percent chance of a cure, I was almost indifferent to the pain and risks that undergoing the surgeries would entail. Because it was certain that I would not live for long if I did not have the surgeries, the potential reward was disproportionately greater than almost any risks.

Now my cancer is growing aggressively and rapidly restricting my life. Through the combined pressure of fluid accumulating in my body cavity and swelling of my liver and spleen, my organs have pushed through my stomach wall and are now protruding beneath my skin. The doctor who saw me at the Urgent Care Center on Sunday and this morning told me that the flexibility of the skin will, however, prevent my organs from bursting through my skin into the open.

This morning, the doctor on call at Urgent Care was once again not able to locate with ultrasound any fluid in a location where it could be removed. Because my CAT scans show that my colon is thickening and ingesting solid food has become too painful to bear, she also had X-rays taken of my colon to see whether it has any blockages. Although my colon is not blocked, I was given a list of symptoms of such blockages to monitor. It is also possible that my intestines could become twisted and blocked as they protrude though hernias in the fascia of the stomach wall. If I develop blockages, the first treatment would entail hospitalizing me with a tube running down through my nose into my stomach.

Because I will die soon and would almost certainly die within a year no matter what treatments I underwent, the suffering and humiliations of further chemotherapy and the risks of prolonging dying loom large in my thinking.

The quality of my life is eroding. Blood tests show that I am anemic. I can't eat normal amounts of normal foods. I can't exercise beyond walks of a dozen blocks or so. The taut skin of my belly is so sensitive that showers are painful, and I dread the mere touch of fabric. I am starting to be afflicted by fevers of unknown origin. Pain is spreading into new locations in my body—today, into my right side, just below the rib cage. Last night, for the first time, I took a painkiller—500 mg of Tylenol—so that I could get comfortable enough to sleep.

This afternoon, Susan and I will meet again with the palliative care specialist, Dr. Glare. I am interested in gaining a better understanding of the risks that I will now be running if I prolong the dying process. I know that my circumstances could change quickly, with dire consequences, and that the converse is not true. For mere mortals, time's arrow has but one direction.

PSYCHOLOGY OF
A DECISION

*His sufferings, growing more and more severe, did their work
and prepared him for death.... Hitherto each individual desire
aroused by suffering or privation, such as hunger, fatigue,
thirst, had brought enjoyment when gratified. But now
privation and suffering were not followed by relief,
and the effort to obtain relief only occasioned fresh suffering.
And so all desires were merged in one—the desire to be rid of all
this pain and from its source, the body.*

—Leo Tolstoy
ANNA KARENINA

NO MORE CHEMOTHERAPY

I have decided to undergo no more chemotherapy. For days, I tried without success to envision my having peace of mind while subjecting myself to a course of irinotecan and Erbitux in the knowledge that, at best, at my advanced stage of incurable disease, this particular chemotherapy's handmaiden, grotesque suffering, would be for the sake of short prolongation of life.

At no time in my internal deliberations was I able to convince myself that this chemotherapy under these circumstances would be good for my family or consistent with my desire to maximize the quality of my remaining days. Unlike my agonizing over the years about countless trivial choices, I was able to make this life-altering decision without angst.

THE BELMONT STAKES

Tomorrow is the Belmont Stakes, the most important day of racing in New York. In the last forty-seven years, I can recall having missed attending only the 1968 running: A man must have priorities.

Because there is no Triple Crown on the line this year—different horses won the Derby and the Preakness this year, and neither of them will contest the Belmont—attendance will be below that of years when a horse is running for the Triple Crown. Nevertheless, the undercard will be loaded with big fields of horses competing for big purses in other races. The race prior to the Belmont Stakes will be the $400,000 Grade I Manhattan, in which the favorite to win will be Christophe's Gio Ponti.

Two races after the Belmont Stakes, we will run Mustang Island, in a modest $48,000 allowance race. Still, it would be a thrill to win a race on Belmont Stakes day.

Because the Belmont Stakes and our race are so close together in time, if I feel well enough, we are going to try to attend that span of races. If we go, we will park our car at Christophe's barn, which would shorten my walk to the grandstand and our box. To maximize my chances of feeling well enough, I will try to take a nap after lunch tomorrow, before we drive out to Belmont Park.

JUNE 6, 2010
SUNDAY 10:09 AM

ON THE RAIL

Several friends have e-mailed to ask if we made it to the races yesterday and to inquire about Mustang Island's degree of success in his race. We did get to Belmont Park in time to see the last four races on the card, including the $400,000 Grade I Manhattan Handicap, the $1,000,000 Grade I Belmont Stakes, and, two races later, Mustang Island's $48,000 allowance race. After the track announcer declared that yesterday's feature race was the 102nd running of the Belmont Stakes, I couldn't resist saying to a friend, "I missed the first fifty-five."

Mustang Island is always anxious to run. We wanted Rajiv Maragh to cover him up early so that he would not burn energy fighting against Maragh's hold on the reins. When a horse is *covered up*, it has a horse directly in front of it and preferably one also just to its outside.

As a result, the horse paces itself to keep from running into another horse. As the race unfolded, as we had anticipated, the horse in the number one post position sprinted away to a clear lead on the extremely firm track surface. Unfortunately, none of the other horses dropped down to the open spot on the rail. Thus, Mustang Island, breaking from the number two post position, had daylight in front of him the whole way. He fought Maragh throughout the race and never relaxed, finishing third, beaten less than two lengths for first and a nose for second. Having lost two races in a row employing this strategy with Mustang Island, perhaps we should permit him to run freely if he wants at the beginning of his next race.

As long as my horses run competitively, adverse racing luck never bothers me. I am mindful that I have won my share of races with an assist from good racing luck.

JUNE 8, 2010
TUESDAY 6:06 PM

WHISTLEJACKET IN MY EYE

When we commenced the project of reshaping the material in this blog as the heart of a book, one of the first subjects that Nanette Stevenson, the book's designer, raised—via conference call on Skype (Nanette lives and works in Alaska)—was artwork for the dust jacket and possibly within the book itself. When she asked if I had any suggestions, I had given no thought about a dust jacket. As we began to discuss suitable art, my eye fell on the stacks of books on my coffee table, and I began fumbling through one, Tamsin Pickeral's *The Horse: 30,000 Years of the Horse in Art*. After flipping through two or three pages, I seized upon a reproduction on page 218 of a 1770 painting by George Stubbs, *A Horse Frightened by a Lion*.

This painting of *A Horse Frightened by a Lion* is part of the permanent collection of the Walker Art Gallery in Liverpool,

England, where Stubbs grew up. Stubbs is known to have produced seventeen paintings on a Romantic horse-versus-lion theme. I saw this particular painting about three and a half years ago at The Frick Collection in New York, when it was one of seventeen Stubbs paintings on various subjects on loan to a traveling exhibition in England and the United States.

Stubbs deliberately portrayed the lions in this series as mangy "demonic mockeries." (That's on page 117 in *Stubbs & the Horse* by Dr. Malcolm Warner, Robin Blake, Lance Mayer, and Gay Myers.) It continues: "In Stubbs' world, the horse is first and noblest among animals and the lion at the other end of the scale.... In his later treatments of the theme, Stubbs made the horse white and the lion more shadowy, heightening the sense of good against evil." To paraphrase the Frick curator's notes, written for a Stubbs exhibition, what mattered is not that the lion would inevitably devour the horse, but rather that the horse struggles nobly.

From the time that I was first exposed to Stubbs, he has been my favorite painter of equine scenes. He was one of the first artists to paint famous thoroughbred racehorses. As a result of gruesome anatomical studies, his understanding of equine anatomy was unmatched by artists in his day. About 1984, I purchased my first art book that prominently featured Stubbs's works, *The Horse in Art* by John Baskett. My small library also contains a copy of Stubbs's 1766 *The Anatomy of the Horse*.

I have never visited the museum that Paul Mellon established in New Haven, the Yale Center for British Art, which houses most of the forty Stubbs works that Mr. Mellon, a leading owner and breeder of thoroughbreds, collected during his lifetime. Nor have I ever actually seen my favorite Stubbs painting, the startlingly modern, life-sized, and almost photo-realistic *Whistlejacket*, acquired in 1997 by the National Gallery in London. I do, however, have *Whistlejacket* forever in my mind's eye.

PSYCHOLOGY OF A DECISION

Although I have always felt profound respect for any individual's urge to postpone his or her death for as long as possible, for many years I have been appalled at the enervating and degrading treatments inflicted on incurable cancer patients. When I was given a chance at a cure, I was cavalier to the point of foolhardiness about the possible risks and probable consequences of undergoing two major abdominal surgeries, each of which included a liver resectioning. I have paid and am paying a terrible price for those unsuccessful surgeries. Nevertheless, even in the midst of my worst postsurgical complications and ordeals, I have never regretted having risked everything for a chance of cure.

The biggest surprise to me when I commenced cancer treatment is that little goes according to plan, and nothing is black or white. Even with as common a cancer as my metastatic colon cancer, a high percentage of my symptoms from the cancer and from the treatments have been inexplicable to even the renowned specialists taking care of me. Their ability to predict the consequences of undertaking a given treatment seemed to me to be limited, not by their undoubted expertise, but by the primitive state of the art.

Although no treatment option was ever presented to me as a choice between quality of life and quantity of life, I did read implicit understanding of my priorities into my oncologist's approval of my decision to postpone further chemotherapy, in the form of FOLFOX, for as long as possible after the unsuccessful surgeries. Even if FOLFOX had proven to be efficacious for me, which it did not, it was never presented to me, and it is by no means clear, that commencing FOLFOX earlier would have made any difference.

I am in awe of the doctors, nurses, and aides who take care of cancer patients. I cannot imagine that I would be able to go to work each day and watch my patients and their families suffer. In my limited experience, a cancer patient trying to weigh a strategic desire for quality of life against an actual treatment option has to think for himself or herself. The cancer industry is built around extending life, even at the price of compromising quality of life.

Now that I am in palliative care, it is clear to me that the ethos in either hospice care or plain palliative care is for the patient to ingest various drugs to deal with pain, blockages, and other symptoms. When the patient becomes sufficiently helpless, the patient is shifted around in a hospital bed, temporarily installed in his or her home, to avoid bedsores. At the end of life, he or she may well be in diapers and in a coma. If do-not-resuscitate orders are in place and are actually observed, the patient is finally allowed to die, in a process that legally can include the withholding of all sustenance including liquids and the administration of painkillers that may have as a side effect, but not objective, the hastening of death.

Even now, I am enjoying shards of life that would not be available to me if I were in the throes of irinotecan/Erbitux. This morning, friends are driving Susan and me to New Haven to view the world's largest Stubbs collection, bequeathed to Yale by Paul Mellon.

Would I live longer had I submitted to irinotecan/Erbitux? Probably, if that chemotherapy had proved to be efficacious for me. In any case, delaying death might be a mixed blessing. My painfully growing belly vexes me constantly. Up until early 2009, I ate robustly and got some three hours of exercise a day, including an hour of walking. I weighed about 165 pounds. Today, I can eat little, and I get very little exercise. Consequently, except in the belly, I am the thinnest that I have ever been in my life. I weigh 180 pounds.

If I could be granted another thirty years of life in my current condition, would I wish for it? No, I would not. It follows that I have no interest in trying to withstand six months of a hideous chemotherapy, like irinotecan/Erbitux, in hopes of postponing the relief of death for several months.

JUNE 14, 2010
MONDAY 9:38 AM

WITNESS TO GREATNESS

Yesterday morning, Susan and I rode with friends in their Maybach to New Haven to see the Stubbs paintings. The backseat of that splendid automobile is so comfortable and the company in the car was so delightful that the drive to and from New York was pain-free. Of the Stubbs works, the *pièce de résistance* in the Yale Center for British Art is a monumental canvas of the first of Stubbs's known portrayals of a lion and horse encounter.

When we returned from New Haven, it started to rain. David and I watched on television the second start by the phenomenal rookie pitcher, Stephen Strasburg. After dinner, on a television broadcast from Los Angeles, we watched Zenyatta remain undefeated and set a new record for thoroughbred horses, by winning her seventeenth straight race.

By the time I went to bed, I had a fever and was in physical distress. But throughout the day I had been a beneficiary of greatheartedness and a witness to greatness.

DRUGS FOR THE DYING

Because I have wanted to avoid side effects—the most common of which is constipation, which can have serious consequences for someone who has had part of his colon removed—and because I have wanted to avoid addling my brain, I have been fighting the temptation to take painkillers. On June 1, I relented enough to take a Tylenol. Since then, I have taken four more. Last night, a Tylenol was not sufficient to mitigate the pain radiating through my body cavity, and I could not sleep. With great misgivings, I started down the one-way street of drugs for the dying.

I took the minimum dose, 5 mg, of oxycodone HCL (Oxy-Contin) that MSKCC had prescribed for me. The pain eased quickly. An hour or so later, I began sleeping for about a half hour to an hour at a time, for probably about three hours of sleep altogether. Oxycodone mimics morphine. My current prescription permits me to take up to 40 mg per day.

WHAT TO DO NOW?

Today, I took two 5-mg doses of OxyContin (oxycodone), equal to one-quarter of the maximum 40-mg daily dose that MSKCC prescribed for me. The drug arrests my pain, but it also has paralyzed what is left of my colon. Having gotten only about three hours of sleep last night plus a nap of about an hour's duration today, I am exhausted. When I go to sleep, I reawaken almost immediately from an urge to have a bowel movement, which my inert colon is incapable of delivering. In the absence of better ideas, I will keep drinking a lot of water, not eat, and try to stay awake. I have a previously scheduled appointment tomorrow afternoon with Dr. Glare at MSKCC.

However well intended, feeding drugs to terminal cancer patients, especially drugs to counteract the side effects of other drugs, strikes me as ofttimes mad. Predicting the side effects of any drugs in people as compromised as incurables who have previously been subjected to various draconian treatments is guesswork, however educated.

JUNE 18, 2010

FRIDAY 5:35 AM

A WORLD OF MADNESS

Because I have expressed my willingness to trade quantity of life for quality of life, MSKCC's Palliative Care Unit's protocol for me yesterday included having me see a staff psychiatrist to make sure that I am not depressed. The psychiatrist agreed with the palliative-care professionals, and with my own assessment, that I show no signs of depression. As I have tried to reassure all concerned, my thinking about quality of life versus quantity of life and my personal disinterest in staving off death to a bedridden, drug-addled bitter end were formed long before I was diagnosed as an incurable.

Yesterday, I continued an extended conversation with the professionals in palliative care not only about what they might be able to do to alleviate some of my symptoms, but also about the limitations on their care in general. (Given the boundaries of laws and medical ethics, I doubt that their limitations are any different from those of professionals in other secular cancer centers.) My understanding is that much of what they can do is to alleviate patients' pain with opiates. The opioids cause constipation, which can cause an anal fissure, as indeed happened to me in the aftermath of all of the complications from my last surgery. If I had to choose between dying sooner and getting another anal fissure, dying sooner would be an easy choice. As the doctors tried to do after my surgeries, the palliative-care doctors and nurses try to offset the opiates with other drugs to avoid constipation, while trying not to trigger diarrhea. Palliative care can also include deep sedation in the final stage of dying, but not before.

I told the professionals in the palliative care group that, to me, taking palliative drugs that would require taking other drugs to treat the side effects of the palliative drugs, and so on, was a form of madness. The nurse practitioner nodded and said that was the mad world in which they worked. I replied that I wanted to avoid entering that world. As I told them—and they tacitly seemed to agree—the only alternative for someone in my situation who does not wish either to enter that world of madness or, out of respect for the sensibilities of family, to commit suicide, is to stop taking food and water once prolonging of the dying process seems unwise or intolerable.

I agreed to go back for a third time to the Urgent Care Center to see if the doctors there or in Interventional Radiology can find a way to put a needle in my body and drain from it some of the steadily mounting fluid. I agreed to try some non-opioids, which supposedly do not cause severe side effects, to try to alleviate some of my symptoms. I read the package insert before I took one of the pills: One of the possible side effects of this non-opioid painkiller is constipation so severe that it requires manual extraction of the feces. I did not take this painkiller.

A DARK COMEDY, IN THREE ACTS

At 7:30 a.m. Saturday morning, Susan and I reported to the Urgent Care Center, as instructed on the computerized appointment schedule that MSKCC maintains for each of its patients. Dr. Glare had tried to arrange for me to bypass the Urgent Care Center and be seen directly by the Interventional Radiology (IR) department, to determine whether the IR doctors could remove some of the ever-increasing fluid in my body cavity. I was feeling grumpy about having to keep a third appointment in Urgent Care, as the first two appointments had proved to be exhausting wastes of time.

We waited about two hours for a confused physician to see me, who said that my records showed that I had gone to the IR department two days previously to have fluid drawn. Nevertheless, she agreed to examine me ultrasonically to see if there was fluid she could access with a needle. She found an accessible pocket and drained some three liters of fluid. I would guess that there are another ten liters where that came from. After I was dismissed and we went home and had lunch, I discovered that the Urgent Care Center had forgotten to disconnect my Mediport. In a lamentable mood, I paid my second—this time gratuitous—trip of the day to Urgent Care.

When I returned home, I worked on this book via Skype with the designer and editor and then suddenly developed chills so severe in the ninety-degree summer temperature that I had to don a heavy cashmere jacket, ski parka, stocking cap, and Arctic gloves to stop shaking. My temperature shot up to 101.9 degrees. In dread of another long, uncomfortable wait, I delayed going back to Urgent Care. I was trying to think through whether dying of an infection might not be as good an alternative as I am likely to get.

I concluded that I don't know anything about dying of an infection and reluctantly went back to Urgent Care about 7:00 p.m. About three hours later, the doctor who saw me recommended that I be hospitalized and begin intravenous antibiotics. In spite of my having had an adverse reaction to a penicillin drug in the past, she recommended using a penicillin-related drug. When I asked what would happen if I got an anaphylactic reaction, she said, "We will intubate you."

After signing a form indicating that I was acting against MSKCC's advice, I got a prescription for non-penicillin antibiotics, and we went home. Altogether, my three trips to the Urgent Care Center had involved about eight hours of waiting around, with pain mounting as fatigue deepened. As I finish writing this post on Father's Day morning, I have gotten a little rest, and I have only a mild fever.

JUNE 21, 2010

MONDAY 7:14 PM

ACHILLES WOULD UNDERSTAND

I'm headed back to the Urgent Care Center. First, we'll have dinner at home. There is no telling how many hours I will have to wait in the emergency room to be seen. The antibiotic, Levaquin, that I began taking on Saturday may have swollen or ruptured my right Achilles tendon.

DANTE'S WAITING ROOM

After studying my blood tests and examining my right Achilles tendon, the physician on duty, Dr. Adam Klotz, thought that the soreness and stiffness in my heel was either a blood clot (I'm not kidding) or a swollen tendon caused, most likely, by Levaquin. Rather than, as he said, torture me with more tests or readmission to the hospital, he sent me home about midnight with two new antibiotics. I was most grateful for Dr. Klotz's common sense and compassion.

Heaven, Hell, and Purgatory can be located anywhere. The emergency room of a cancer center is an outpost of Purgatory.

STILL SOME WORK TO BE DONE

During my adult life, I enjoyed exceptionally good health. Now I find being ill, especially in the context of terminal decline, to be frustratingly inefficient. I have spent the last twelve hours or so, starting about 5:00 p.m. yesterday, dealing with a low-grade fever, bodily pain, diarrhea, and slight nausea. I have been able to eat only small portions of food and sleep only in stretches of about half an hour. No interactions with people outside my immediate household, no writing, no reading, only small snatches of television.

An existence consisting primarily of tending to my miseries while in inexorable decline is not only boring, but also strikes me as self-indulgent and ignominious. I am fortunate that, so far, I have still been able to work most days with my colleagues—via e-mail, Skype, and FedEx—in recasting this blog into a book. Otherwise, I would have little work in my waning days.

JUNE 25, 2010
FRIDAY 4:38 PM

RETURNING TO THE WINNER'S CIRCLE

In today's sixth race at Belmont Park, Hot Money circled the field on the turn for home in a seven-furlong turf race and won going away by a half a length under jockey Rajiv Maragh. Unlike today, Hot Money had not gotten smooth trips in his last two races. But those two losses did not deter the bettors from backing him down today to second choice at 2.4-to-1 odds. Hot Money's cumulative gross earnings are now some $87,000. Although I was not well enough to go out to Belmont Park, Susan represented us well in the winner's circle.

Because Hot Money works extremely well on dirt in the mornings, we want to race him once on dirt. Meanwhile, there should be races in coming days for Mustang Island and for Backslider, who would be making his debut.

JUNE 26, 2010
SATURDAY 10:53 AM

TAKING STOCK

With the aid of a drug, Reglan, I can still eat small portions of chicken, potatoes, rice, pasta, eggs, cream of wheat cereal, yogurt, molasses cookies, applesauce, and cooked zucchini. I can drink water, Gatorade, and cranberry juice. I can sleep, in spite of a worsening skin rash, in the fetal position on my left side for a half hour at a time, occasionally even an hour, before I am awakened by musculoskeletal pain from lying in one position or by the need for a bowel movement. I can see well enough by peering to hunt-and-peck out this post, to read in good lighting, and to watch television. I can concentrate well enough to read a newspaper article, though not a book. I can still work on the book with our editorial and design team for brief, intense stretches. I can still remember things, though sometimes I have real lapses in short-term memory—I am hoping such lapses are owing to the Reglan. I enjoy listening to music, mainly classical and jazz. I am strong enough to walk a few blocks, but I would have physical difficulty enduring another long wait in the Urgent Care Center. I can wear a loose polo shirt, though not a tee shirt or a dress shirt. In certain positions, if I am not too tired, I can sit pain-free in a comfortable chair for a couple of hours. Sometimes I have no fever, often I have only a mild fever, and only rarely do I have chills and shakes. I am taking Tylenol, but only when my temperature starts rising.

We interviewed home hospice services late last week and expect to sign a contract with one early this coming week.

SEARCH FOR
REMAINING MEANING

*The meaning of the greatest secret that human poetry
and human thought and belief have to impart:
The salvation of man is through love and in love.*

—Viktor E. Frankl
MAN'S SEARCH FOR MEANING

TO SLEEP, PERCHANCE TO DREAM

I have been taking about a half dozen naps each day. Most of them are short—often of only a few minutes duration. Invariably, I dream. These dreams are rarely nightmares.

I had long thought that it took awhile to go through several stages of sleep before reaching the rapid eye movement (REM) stage that I had understood was a prerequisite for dreaming. I suppose that I must have a cumulative deficit not only of sleep in general, but of REM sleep in particular.

TIME AND AGAIN

As the Fourth of July approaches, I think nostalgically of our many Fourths of July on Long Island playing in the annual three-day Liberty Bell golf tournament and celebrating with friends at their lake place in Connecticut.

I think of my life as having taken place before and after diagnosis. As I look back today, the sixteen months since diagnosis seem almost as long as the sixty-six years preceding it. Yet, since diagnosis, I have celebrated only one round of birthdays for my wife, two children, mother, father, and myself, as well as one wedding anniversary, two Memorial Days, and one set of all other national holidays. We will celebrate mother's ninety-sixth birthday on July 16.

SEARCH FOR REMAINING MEANING

On Friday, July 1, the Interventional Radiology surgeons at MSKCC drained three liters of fluid from, and inserted a Tenckhoff catheter into, my peritoneal cavity. Because the catheter sticks out where the belt line of trousers would be, I am wearing sweatpants. I have a team of three "family substitute caregivers." Each morning and evening, one of them removes another half liter of fluid from me, for an allowed daily maximum of one liter per day. Although I have had fevers since Saturday, I feel less pressure on my stomach and can eat and sleep more—so, on balance, I feel better.

I have reached the point that, whenever I leave the hospital, I feel as if a piece of me remains behind. Although I have entered home hospice care, MSKCC will remain my primary medical resource. Nevertheless, a primary objective of home hospice care is to minimize hospital visits.

With so much of each of my days and nights now consumed with the ignominy of self-maintenance, it is difficult for me to find adequate meaning to justify my continued existence in anything other than spending as much time as possible doing things with my immediate family.

Today seems a logical time to end my blogging. I have not yet succumbed to the blandishments of the doctors, nurse practitioners, nurses, and hospice workers to take opioids.

I am still myself.

SUSAN'S POSTSCRIPT

❧

I am not sure why Charlie decided to stop blogging in early July. He did not seem too weak to continue, and the writing had proven itself to be cathartic for him. Did he think he had only a few days left? Was he too focused on his care—and being vigilant that it did not violate any of his self-imposed guidelines—to take the time or effort? Did he think he had said all he had to say? Did he think the final days would be too grim to report? His reasons were never clear to me, and he was vague in answering my questions.

I never understood why or how the blog evolved as it did. The initial idea was only to keep family and friends updated on his condition. Charlie was not just a "private person." He was some-one who never revealed his true feelings to anyone but me and, often, was rather secretive about how he felt. I am sure his good friends had noticed over the years how he deflected a conversation that was becoming too personal for his comfort. I was surprised to see him revealing his innermost thoughts in the blog. As the posts continued, the tables were turned, and I was the one who was uncomfortable.

When he was writing, Charlie would typically spend two or three hours on each blog entry, sometimes more if he needed to research something. (He did not have all those facts, like the details of his grandfather's life, at his fingertips.) He sometimes changed an entry a day or two later, after thinking about it or hearing criticism from David or me. We tried to suggest to him that the nature of blogging does not include after-the-fact editing, that it is informal and, by definition, imperfect. But "imperfect" was not part of Charlie's makeup. During his illness, he would often get up in the middle of the night to sit in the living-room chair and edit his last blog entry.

Charlie did not like to use his laptop on his lap, so he rested it on top of a giant dictionary, perched on a small table that he had moved in front of a living-room chair. Books on death and dying surrounded him. I became accustomed to these books as part of the decor, but I am sure they were strange and perhaps upsetting to visitors. (My housekeeper complained about them regularly.) Most of these books had been ordered using my Amazon account, so Amazon has taken to welcoming me with suggestions about more death books I might like to buy.

A cliché regarding couples who have been together a long time—we met in 1967—suggests that they can and do finish each other's sentences. Despite our closeness and many shared values and interests, Charlie and I were never like that. We often had two versions of the same story or what was likely to happen, especially when it came to health issues. Charlie conjured up the worst-case scenario, perhaps so he could be favorably surprised and never disappointed.

During the years of our marriage, there was a suspected broken leg that turned out to be shinsplints, a possible case of pneumonia that turned out to be a bad cold, heart attack symptoms that were caused by hyperventilation, and many more. By contrast, I am a perennial optimist who rarely, if ever, calls the doctor, on the assumption that whatever I have is minor and will go away. Although I saw humor in some of Charlie's health worries, the differences in the way we approached illnesses or suspected illnesses could be a source of friction from time to time.

Given Charlie's disposition, the way he handled his cancer diagnosis and the months of his illness was remarkable. To be sure, he had moments of anxiety over suspected side effects of his medications, and some of the trips to the Urgent Care Center were unnecessary. When it came to the big picture, there was no panic. Instead, he was serene and coldly analytical, perhaps because the worst-case scenario had happened, and he was more mentally prepared than I would have been.

I am not surprised that Charlie adopted Victor Frankl's suggestion that, with death at hand, the individual needs to make his or her own choices, as a way of maintaining dignity and freedom. Charlie also made a conscious decision not to waste his time being depressed. But deciding and doing are two different things. I am filled with admiration that Charlie was able to be a "rational optimist," even in the face of his death.

In writing about events between early July and September 30, when Charlie died, I may not report exactly as he would have, but I will try to give you my picture of what happened. It was a sad and difficult time, but one that included much love and, for Charlie, great progress in his search for meaning in his life, which consumed so much of his thinking during the last year.

Let me begin with the physical aspects of Charlie's final three months. The period began as badly as it ended. In mid-March, while we were in New Zealand, Charlie's belly began to grow. We thought he had been overeating, although it was odd that, for the first time in his life, he seemed to put on extra pounds around his middle. By June, his belly was huge. The explanation was ascites associated with the cancer. Ascites is a buildup of fluid in the peritoneal cavity. Just before he stopped blogging in July, a catheter that could be drained at home was inserted in his belly. Over time we removed more than twenty-five liters of fluid using the catheter, so it is easy to imagine how uncomfortable he was.

Over the summer, when Charlie decided to enter hospice care, I became more aware of its holistic approach, and the teachings of the hospice movement that mental peace is just as important as physical comfort. The case manager for Calvary Hospice@Home, Dan Doniger, emphasized that philosophy during the three months Charlie was his patient. His guidance had a positive influence on Charlie's peace of mind. Dan was endlessly persevering in finding and offering physical comfort solutions that did not conflict with Charlie's wish to maintain dignity and avoid hospitalization.

With the use of the catheter and some unknown natural processes, Charlie's belly became smaller. But the cancer was making it difficult for him to eat more than a small meal and eventually to eat anything at all. Various strategies intended to allow Charlie to tolerate eating failed, including medication that helped food move through his digestive tract. Pain medications were available, although Charlie feared their side effects. It took a while for Dan to win his confidence, but after he did, Charlie would accept Dan's assurances that side effects could be avoided or balanced with proper care and procedures.

About the middle of August, Charlie stopped eating all but a few bits of food, because anything he ate brought him discomfort. He stopped eating completely on August 18. His appetite was poor, so he did not find this difficult. Dan warned me that there would likely be serious deterioration in Charlie's condition during the following weekend, and I was given instructions on how to handle various situations. Dan consulted with Sue Derby, the palliative care nurse practitioner from MSKCC, a psychiatrist, and the medical director of Calvary Hospice, and they sent medications to treat likely symptoms during the dying stage, should they be needed. They wanted me to be prepared to keep Charlie comfortable at home.

According to Dan, one thing was immediately probable: a decline in lucidity. As the weekend went along, however, there was no change in Charlie's condition. He went on reading the newspaper, checking his Blackberry, writing to his editors, talking with David and me, following his racehorses, and watching football.

One of the harder issues that we dealt with in late summer was Dan's impending vacation. Dan had told me, several weeks earlier, that he was going away on September 5. We decided together not to tell Charlie, because he was unlikely to live that long, and because, were he still alive, he might be alarmed to spend two weeks with a substitute nurse and case manager.

On September 2, a day when Charlie was up and about the house, and taking short walks in the neighborhood, we told him about Dan's vacation. He took the news calmly but told me later in the afternoon his suspicions that Dan was taking himself "off the case." Charlie speculated that Dan was fed up with his many protestations about taking the medications Dan wanted him to take. I convinced Charlie that this was not correct and that we had not wanted needlessly to upset him. The substitute nurse, Deb Lau, was introduced. Dan gave Charlie what we expected to be a final good-bye.

When Dan returned in two weeks, Charlie was still following his daily routine: He would get up in the morning, dress in a sport coat and golf shirt, read the paper in the living room, listen to classical music, nap on and off through the day, and watch a movie with me in the evening. Dan and all the experts and hospice workers were amazed that he was not just alive, but so clearheaded.

Up in my office, which is next to our library and television room, I did some online research and learned that Charlie was living far longer than expected for a person with cancer who is not eating. Later, I heard that his longevity without eating prompted wonderment and an invitation to Dan from Sue Derby to speak about Charlie to the palliative care team at MSKCC.

During the last week in September, there were lapses in Charlie's mental clarity, but he did not lose significant lucidity until two days before he died. The day before he died, he walked himself into the bathroom, although, for the first time, he had an episode of disorientation and spent the day in bed. Much of the time, he appeared to be sleeping with his eyes open. But when our granddaughter, Kayla, called him on the phone, he responded to what she was saying with short phrases like, "That must have been fun" or "That's great."

During the night of September 29, the hospice workers told David and me that Charlie did not have much time left, but that I should go to bed. They would wake me if his death were imminent. During this time, Charlie's pain was well controlled.

On the morning of September 30, Charlie was sleeping with his eyes open. The person working the night shift, Nina Museliani, whom we hired to care for Charlie and carry out the plan set up by the hospice, did not leave when her shift was over. She and the day-shift person, Jason Cepeda, worked together to make sure that Charlie was comfortable. (These two and several others we hired had been like supportive family to Charlie during the last part of his life.) Around noon, one of them noticed that his feet were getting cold, one of many signs that his system was shutting down. The hospice team left the bedroom. David and I took up positions on either side of him, holding his hands or gently rubbing his shoulders. He was responsive to what we were saying and, at least once, smiled at me. About two hours later, he died peacefully. He weighed 109 pounds, but in many other ways he was still "himself."

Charlie wrote in his blog that Dr. Sherwin Nuland, the author of *How We Die*, admitted in an epilogue to his book that even he—knowing so much about dying and having observed so much in his long career as a surgeon—might be likely to show indecision that could compound the pain and agony of his final days. Charlie feared that he might do the same, even having read extensively about death and deciding early on that he wanted quality rather than quantity of life. Charlie would never show that indecision. There were no crises, such as difficulty in breathing, to cause panic.

He died comfortably at home, in his own bed, with two members of his family at his side. By his choice, he was not heavily medicated, but I am certain that he did not suffer.

Dan and those at Charlie's bedside, like others in MSKCC's palliative care, taught me that dying is a part of life that can have special quality. Two weeks before he died, when Charlie told us in the immediate family that he had never felt so loved in his life, I understood what they meant. Cancer is a terrible disease, but it does allow a final chance to share love, confessions, regrets, joys, and intimacy that those succumbing to sudden deaths and their families do not have.

Charlie's emotional well-being in his final days was bolstered by the astonishing number of e-mails, letters, and cards that he received and by the depth and sincerity that the messages contained. After I wrote in my blog about the pleasure Charlie got from these messages and how appropriate it seemed to share these thoughts with Charlie—what he had meant to the people in his life—a new flood of e-mails arrived.

The sincerity and personal nature of these communications was wonderful. It is hard to exaggerate the importance these letters had in helping Charlie achieve mental peace at the end of a difficult road. Perhaps by revealing so much of himself in the blog, Charlie allowed his friends, colleagues, and family the freedom to do the same.

In the final weeks, Charlie and I spent a great deal of time talking, recalling good times, thanking each other for ways that we each contributed to the happiness in our family, telling how much we loved each other, and sharing our fears. I reassured Charlie that I would miss him terribly, but that I would be fine with the support of family and friends. At the end, I let him know that it was all right for him to let go, and we said our good-byes.

December 1, 2010, New York City

THE POSTS

A POSSIBILITY

IN THE RECOVERY ROOM

SARATOGA SPRINGS

BACK IN ROOM 1937

ON THE INSIDE

WHAT IS REQUIRED OF ME NOW?

PSYCHOLOGY OF A DECISION

SEARCH FOR REMAINING MEANING

A Note on George Stubbs
(1724 – 1806)

A Horse Frightened by a Lion
Oil on Canvas (1770)
Walker Art Gallery, National Museums Liverpool, England

Horse Attacked by a Lion
Enamel on Copper (1769)
Tate Gallery, London

George Stubbs's elegant depictions of the early thoroughbred horse along with his meticulous drawings in *The Anatomy of the Horse* rank him as one of the foremost painters of horses and horse portraiture. *A Horse Frightened by a Lion* (pictured on the jacket) shows the shock of a horse in the Creswell Crags wilds of England upon realizing that a lion is stalking him. *Horse Attacked by a Lion* (pictured in black and white at the end of the text) shows a horse in the agony of being savaged by a lion.

About the Author

Charles Harris (1943–2010) was born in Greensboro, North Carolina, and grew up in Jacksonville, Florida. He graduated from the Hill School in Pennsylvania, Princeton University (BA in English, 1964), and Columbia University (MBA in Finance, 1967). From 1966 to 1983, he worked on Wall Street, first as a securities analyst and later as chairman of an investment advisory firm.

In 1983, he became an entrepreneur, later creating Harris & Harris Group as a publicly traded venture capital firm that invests in nanotechnology, from which he retired in 2008 as chairman and chief executive officer.

He was a trustee of The Institute for Genomic Research, of Nidus, and, for two terms, of Cold Spring Harbor Laboratory. He was a Life Sustaining Fellow of the Massachusetts Institute of Technology.

Charles Harris lived in New York City, where he and his wife, Susan, raised their two children, Elizabeth and David. He bred and raced thoroughbred horses for thirty-six years. He began writing a blog to keep his family and friends current with his cancer treatments. Those blog posts evolved into *Incurable*, his first book. He died September 30, 2010.